We Alone

We Alone

How Humans
Have Conquered
the Planet and
Can Also Save It

David Western

Yale UNIVERSITY PRESS
New Haven & London

Published with assistance from the foundation established in memory of Henry Weldon Barnes of the Class of 1882, Yale College.

Yale University Press books may be purchased in quantity for educational, business, or promotional use. For information, please e-mail sales.press@yale.edu (U.S. office) or sales@yaleup.co.uk (U.K. office).

Set in ITC Galliard Pro and Gotham type by IDS Infotech, Ltd.
Printed in the United States of America.

Library of Congress Control Number: 2020935060
ISBN 978-0-300-25116-6 (hardcover : alk. paper)

A catalogue record for this book is available from the British Library.

This paper meets the requirements of ANSI/NISO Z39.48-1992 (Permanence of Paper).

10 9 8 7 6 5 4 3 2 1

All photographs are from the author's collection unless otherwise noted.

To Shirley, Carissa, and Guy,
and the communities that have done so much to
conserve wildlife

Contents

Part II. The Human Age

Part III. Our Once and Future Planet

Illustrations follow page 138

Preface

We stand at a pivotal point in human history. In our rise from small, scattered Neolithic communities living precariously, we have become so supremely dominant as to reshape nature, change the course of evolution, and engineer a new geological age, the Anthropocene. In the process we have created a global economy that has narrowed our food webs and stretched our supply chains to the point that we can no longer sense or contain our planetary impact.

Belatedly, climate warming has risen to the top of the international agenda as hotter summers, colder winters, stronger hurricanes, torrential floods, searing droughts, and rising sea levels impinge on our daily lives. Trapped between a receding industrial age powered by toxic and dwindling fossil fuels and the fourth industrial revolution and circular economy promising hope of a greener planet and sustainable lifestyles, we face a tragedy of the global commons for lack of action.

Prescriptions for a sustainable future range from strong central government control to trusting in the invisible hand of free markets and reliance on rational choices. Yet neither Big Government nor Free Markets have yet solved the ultimate human challenge of living within planetary limits.

In *We Alone* I look for answers by delving into how we rose from a lowly savanna primate to conquer the Earth and examine how successful societies avoided the pitfalls of overuse and social breakdown. My exploration is partly a personal voyage tracing my evolution from hunter to conservationist, highlighting insights I've gleaned from observing communities:

from the Maasai surviving droughts to Californians up against intensifying droughts and wildfires caused by global warming. I also draw on scientific discoveries over the last half century to show how we humans are far from being constrained by the selfish gene and limited by local ecology; instead our success lies in cooperation and cultural institutions that enable us to create novel economies and lifestyles that defy the biological imperative to reproduce to the limits of food supply.

I argue that our global conquest lay in breaking biological barriers, domesticating the selfish gene, and curbing the downsides of our actions for larger common gains. The more ecologically emancipated we became, the greater grew our ability to shift beyond conserving food and water for survival to saving whales, art, music, and cultural traditions, based on our newfound knowledge and sensibilities. Conserving other species lifts us to the highest plane of altruism, beyond kinship, tribe, and economic self-interest.

Our future well-being depends on our unique capacity for cooperation, foresight, and planning as well as on new technologies and green economies, rather than in using a vanishing Pleistocene Age as a template. No less than in the past, our success hinges on using our emotions, morality, and expanded empathy to create the world we wish for rather than the polluted and degraded planet we have inherited.

We Alone takes up where Aldo Leopold's *A Sand County Almanac* leaves off by showing that we can scale up from husbanding the land to sustaining the planet within boundary limits. Neither the end-of-nature pessimists nor rational optimists offer solutions for cleaning up the global problems we have created. Our future lies in the collective actions of billions of citizens, not in philosophical debates and scientific prescriptions.

We Alone is written for a popular audience, but it is also intended to appeal to students looking for answers to who we are and how we can become good custodians of the global commons.

We Alone

Introduction: Confronting the Human Age

This book could be a depressing one about the end of our planet because of human folly. It isn't. Nor is it about wildlife and conservation in Africa, although I lay the foundation for a far larger exploration of our past and future in southern Kenya, where I have worked for the past five decades. Instead, this book is about another savanna species altogether: us.

In *We Alone* I look at conservation from the dual perspective of our ancient savanna origins, which shaped our nature, and the new global age in which we are reshaping nature. What is so unique about our species as to have propelled us from a small-time savanna primate to a globally superdominant species capable of changing nature, ourselves, and our planet? What lessons do cultures that beat the ecological odds and thrived for millennia have to offer our global society, now knocking against planetary limits?

My exploration began with the role conservation played in the survival of Maasai pastoralists battling droughts and competing with wildlife. If they had to struggle so hard to combat the elements, how did our early ancestors, the puniest of the great apes, outcompete all other species and rise to global supremacy? My interests grew with visits to other traditional societies, among them !Kung hunter-gatherers of the Kalahari, Mbuti pygmies of the Ituri Forest, Konso and Gammo farmers of the Ethiopian Highlands, Bedouins in the Middle East and Sahara, Andean farmers of Peru, and yak herders of the Tibetan Plateau.

The exploration led me on a yet deeper quest into the meaning of conservation itself and how it evolved from its roots in survival and necessity to encompass the bewildering variety of things we conserve today—from whales, lions, and wolves to historical buildings, art, music, and above all the diversity of life itself. What, if anything, do these varied strands of conservation have in common?

Over the course of barely two centuries, subsistence herding and farming societies tied to rainfall and the seasons have coalesced into a global society and interwoven economy. Amboseli, situated beneath the rising mass of Kilimanjaro in Kenya, gives a snapshot of the last vestiges of the Neolithic Age merging into the Anthropocene—a world in transition from nature shaping our livelihoods and culture to one in which we are molding nature for our own safety, comfort, and enjoyment—often with unknown and unintended consequences. The Anthropocene marks an extraordinary juncture in human history when thousands of years of differentiation in cultures, lifestyles, and languages are converging into a single entangled community.

In the early 1970s my first visit to California, half a world away from Amboseli, jolted all my expectations of America's third largest metropolis, Los Angeles. My taxi from the airport ground to a standstill, blocked by an endless stream of traffic filling the Santa Monica Freeway bumper to bumper in both directions. An acrid blanket of urban smog stung my eyes, burned my lungs, and hid the iconic Hollywood sign up in the hills. What good do riches do, I wondered, if the price is clogged freeways and congested lungs? On university campuses where I had come to lecture on African wildlife, I was struck by the irony of Americans' interest in African animals—many traveling to Africa to film lions, elephants, and the great Serengeti wildebeest migrations—while they did little to save their own: the bald eagle, the national symbol; the last of the country's wolves; the endangered grizzly; and the plains bison.

Nearly half a century later pollution levels in Los Angeles and other American cities are down 90 percent—despite worse traffic jams, a surging population, and a twentyfold increase in gross domestic product. The bald eagle has made a comeback, the grizzly has been delisted as an endangered species, the wolf has been reintroduced to Yellowstone and is recolonizing the American Northwest, and the plains bison population has grown to over half a million from the mere few hundred that survived the greatest wildlife slaughter in history. California has become a hub for new high-tech industries, leading America in the shift from fossil fuels to renewable energy.

Most of the world has followed America's fossil fuel path to industrialization. China, drawing on its vast coal reserves, is nearing America's economic might, and Asia is following in the fast lane to development. Bent on catching up with the West, the rest of the world is tracking the well-trodden path of pell-mell growth, confident that the damage can be fixed later. But with populations and economies in the developing world growing much faster than they did in the West at peak industrialization, the cost to human health and the environment is now far too great to fix later. Dense smog blanketing Beijing, Shanghai, and other major cities accounts for nearly a third of all deaths in Mainland China, almost as many as smoking. Beijing residents are no longer prepared to defer the costs when it means wearing gas masks to work and keeping their children out of school for days on end. China now deems the cost so great it has spent $350 billion in remediation and is investing heavily in becoming a world leader in renewable energy sources.

For all the progress the United States has made, its victory over pollution is far from won. To the contrary, it faces a far greater threat than ever from the by-products of its wealth heating up the atmosphere and oceans and destabilizing the global climate and hydrological and nutrient cycles. The greenhouse gases caused by burning fossil fuels powering industrialization have raised the Earth's thermostat, causing a cascade of changes seen in melting polar ice caps, receding glaciers, rising sea levels, and acidifying oceans. Summers are getting hotter, blizzards fiercer, hurricanes stronger, floods more frequent, and droughts harsher. In California, rising temperatures have deepened a decade-long drought, thinned winter snowpacks, evaporated reservoirs, dropped water tables to all-time lows, and taken a heavy toll on the Golden State's fruit, nut, and vegetable farms supplying homes across America. Parched vegetation has sparked fierce wildfires, fueled by high winds, that flare up earlier in the season, burn hotter, persist longer, and destroy thousands of homes across the state. In December 2017, two hundred thousand people were evacuated from Ventura County north of Los Angeles as firefighters battled raging fires driven by the strongest Santa Ana winds on record. In 2018 California suffered the deadliest fires in its history. More than eighty-five hundred fires, driven by searing winds, killed eighty-six people, burned 1.9 million acres of land, destroyed eighteen thousand properties, and caused $14.5 billion in damages.

Is it possible to achieve a wealthy, healthy life without overusing resources and sullying our planet? Oddly, megacities like Los Angeles, New York, Mumbai, Lagos, Nairobi, Shanghai, and Beijing are, for better or worse, our new frontier and the best hope for saving our planet. Cities are the origins of our civilizations, the centers of power, the locus of the modern industrial states, and the epicenter of innovations. Despite their higher crime rates and stress, cities are the magnets for modern economies and communities seeking a richer life or escaping vanishing rural livelihoods.

In the new ecosystems we've created, achieving a healthy and satisfying life and saving other species will depend on re-creating the conditions that made us a superdominant species. We must draw on our emotions, morality, and empathy no less than on our rational mind and technical skills to forge universal agreements on our common future. I conclude that our unique human ability for large-scale cooperation and empathy can save nature—and ourselves.

I

The Roots of Our Success

1

Survival in the Savannas

I was so consumed by the famine and suffering among the Maasai in Amboseli during the searing drought of the 1970s I almost missed a puzzling anomaly that would shape my work in the coming decades: how, faced with collapse, did the Maasai pull through in competition with superbly adapted wildlife? Come to that, how did our puny ancestors emerge from our savanna birthplace to reach such supreme dominance as to change the face of the Earth and threaten its very future? More puzzling yet, why, at the pinnacle of our ecological success, did we begin to conserve some of the very species we had conquered?

Amboseli, at the time a 3,370-square-kilometer game reserve stretched along the Kenya border with Tanzania, captures the essence of the African savannas, and is an ideal exemplar of the story of humans' rise from lowly primate to global dominance. To the south, the land descends from the snow-capped peak of Kilimanjaro through alpine meadows, moorlands, rain forest, savanna woodlands, and scattered scrub to wide treeless plain. To the west, 150 kilometers across the Great Rift Valley, lies Serengeti National Park. To the north, barely visible on the horizon, rise the foothills of the Kenya Highlands. Sixty kilometers to the east a blue ridgeline of young volcanic hills marks the Chyulu Range and the boundary of Tsavo West National Park.

The young Scottish explorer Joseph Thomson was captivated by the enigma of Amboseli when he was the first European to traverse Maasailand. He wrote in his 1885 account of the journey, *Through Maasai Land:*

"Conceive of yourself standing in the centre of the plain. In your immediate vicinity there is not a blade of grass to relieve the barren aspect of the damp muddy sand, which, impregnated with various salts, is unfavorable to the growth of any vegetation. Here and there, however, in the horizon are to be detected a few sheets of water, surrounded by rings of green grass, and a few straggling trees or scrubby bushes. In spite of the desolate and barren aspect of the country, game is seen in marvelous abundance. The question [that] naturally rises in one's mind is, how can such enormous numbers of large game live in this extraordinary desert?"[1]

From where I first pitched my tent in 1967 to start my research and conservation work, I looked out on the same scene that had so enchanted Joseph Thomson. Thousands of wildebeest, gazelle, impala, hartebeest, buffalo, giraffe, elephant, and rhino were tightly packed on the edge of the plains and in the elegant fever tree woodlands. Like Thomson, I was intrigued by the spectacular abundance of animals living in such parched country. However, another, more baffling enigma was uppermost in my mind as I watched a savanna scene expunged from the lyrical portraits of Amboseli—the large herds of cattle threading their way through the wild-life concentrations. How could so many Maasai livestock live in seeming harmony with wildlife?

A run of good years had restocked the herds after a harsh drought in 1961 killed off most Maasai cattle and large numbers of wildlife. By 1967 the wet years had run their course, leaving tall stands of dry grass rippling across the open plains. By the early 1970s wildlife and livestock began losing bodily condition as the grasses withered and high winds drove billowing dust clouds across the flats. By 1973 I recorded the first calves dying around the settlements and took note of herders driving their emaciated cattle miles from water in search of pasture.

Month after month I monitored the pasture and animals as the grasses shrank to a wiry stubble, expecting a mounting body count. Oddly, despite the cattle looking like walking cadavers, few died. How could such emaciated animals survive the long grueling treks and irregular watering in a landscape hauntingly reminiscent of the American Dust Bowl of the 1930s?

When I first came to Amboseli, the Maasai, archetypal red-robed, spear-toting cattle herders of the East African savannas, were portrayed as unbending traditionalists preoccupied with the prestige of having large herds rather than the quality of their animals. Government livestock officers viewed Maasai cattle, sheep, and goats as ill-adapted to the African savannas and as destructive to the land as locusts. But if the Maasai were such poor herders and their animals so destructive, how did they survive harsh droughts, and why was Amboseli's wildlife still considered second to none?

Graduate biologists from Europe and America were flocking to East Africa in the 1960s to study its national parks as models of natural ecosystems. Fossils found at Olduvai Gorge on the edge of Serengeti in Tanzania 150 kilometers to the southwest, testifying to the coexistence of humans and wildlife for hundreds of thousands of years, did little to dampen the enthusiasm for the African savannas as an ecological tabula rasa. The eviction of pastoralists from Serengeti, Mara, and other parks two decades earlier did nothing to dispel the illusion of pristine Africa.

Having grown up in Tanzania in the pre-park days, I held a different view. Surely there must be a reason why the best wildlife areas were found in the Maasai heartlands. Amboseli proved the ideal place to look for answers. Here, in a game reserve run by the Kajiado County Council, livestock and wildlife still migrated freely with the seasons. I had no idea what to expect but had a hunch that researching the Maasai and wildlife together would explain their coexistence and, perhaps, point to an alternative to evicting them from their lands in the name of conservation. My zoology professor at the University of Nairobi was dead set against my doctoral study when I briefed him on my research plans. Zoology is the study of animals, not people, he told me. If you want to study people, you should enroll in the geography department.

I ignored the advice after getting strong support from my thesis supervisor, Bristol Foster, a rangy iconoclastic Canadian, happier in the field than teaching classes. He suggested I conduct an ecological survey of the Amboseli Game Reserve. The fever trees are dying and the wildlife disappearing, he told me. Conservationists are blaming the destruction on overgrazing and pushing the government of Kenya to evict the Maasai and create a five-hundred-square-kilometer national park in the center of the reserve.

Tracking the migrations in my dilapidated Land Rover was laborious and spine jangling. Plowing through thick bush and often bogged down overnight, I glimpsed only a hazy picture of the shifting herds in the tangled bush country north of Kilimanjaro. Unlike the continually moving massed herds of Serengeti, the herds of wildebeest and zebra in Amboseli were a will-o'-the-wisp, always elusive. I had no way of keeping up with their erratic movements until I bought a forty-year-old Cessna-180 in 1974. Once I was airborne, my surveys shrank from punishing days bashing through bush to an hour's flight and a vulture's-eye view of the migrations.

Baffled by the uncanny skill of the Maasai to locate the best grazing grounds, often beating zebra and wildebeest to small pockets of greenery, I hit on testing their skills by flying herders over the migratory paths. Clad in red robes and carrying cattle switches, my novice passengers were so startled and enthralled with flying that I got nothing but nervous squeals and shouts of delight at first as we banked and dipped over their herds, and raucous shouts as they yelled at their families and friends peering up from the ground.

Once accustomed to flying, the herders pointed out cues they used on migrations. I was no longer the pilot but an apprentice herder following the direction of their cattle switches as they twisted this way and that, exploring the land as adroitly as if the herders were standing on a hilltop. These are the best grazing grounds, they would point out with their herding sticks, and there the best route to avoid elephants and rhinos in thick bush. That old corral over there would make a good temporary settlement, they added, firing details faster than I could take them in.

As adept as they proved to be at gauging the best migratory routes from the air, Maasai herders don't use planes. So how do they figure out when and where to move in the sprawling bush country? I had often flown over young warriors remote from any settlement and wondered at their wanderings. They are *eleenore,* my airborne guides told me: scouts who gather information on pasture conditions, water sources, wildlife—whatever information their families need to decide where and when to move their herds. Information exchanged among individuals and in meetings builds up the shared body of knowledge, *enkingwana,* on which Maasai husbandry and culture is founded and which is passed down through successive generations.

From the monthly flow of my aerial and ground counts and the rich vein of information I tapped from the Maasai social network, the pattern of migrations finally fell into place. The cattle of migratory families stay in better condition, produce more milk, and have healthier calves than the cattle of families that stay put as conditions worsen. Migratory animals grow faster, mature earlier, and are less susceptible to disease and predators than resident herds. The maxim that the early bird catches the worm applies to Maasai livestock: the astute herder stays in the vanguard of the migrations to get a head start on the richest pastures.

The survival and productivity of the herd are the bedrock of Maasai success. Livestock holdings govern how many wives with a bride price a herder can afford. A herder must have a refined knowledge of the savannas and expert husbandry skills to keep his herd safe and healthy and his family supplied with milk and meat. Children add extra hands for splitting the herds and grazing them more efficiently in the dry season. The cow's udder is a barometer of a herder's success. How he grazes his animals, how far he walks them, how often he rests, shades, and waters them are among the myriad of decisions that affect how much milk his wife collects in her gourd to feed the family.

I was fortunate to befriend Parashino Ole Purdul, a powerfully built young elder in his early thirties considered a skilled herder by his age mates. After I gained his trust, assuring him I was in Amboseli to learn about wildlife and Maasai, not as a government officer out to grab land for a national park, Parashino offered me cattle of my own. If I wanted to understand the Maasai, I must see Amboseli through the eyes of a cow, he insisted.[2] Not one to be held back by scientific detachment or bothered by conservation eyebrows raised by my herding cows through a game reserve, I accepted.

Parashino gave me two cows for my apprenticeship: Sotwa, a flawless white animal, and coal-black Matingab. When we were out herding on all-day treks, Parashino instructed me on how to keep my animals healthy and productive. He tutored me on pasture conditions, seasonal changes, watering regimes, diseases, and how to combat heat, thirst, cold, and starvation. I noted his ever-present eye on wildlife—how he navigated his herds around elephants and rhinos, checked out thickets for lions, and kept his animals from straying out of sight. The lead animal and a few others carried clanging bells fashioned of tin or wood to allow Parashino to keep track of the herd in thick bush. Lions will pick off stray animals, he cautioned, but they fear Maasai. I verified his claim one day when I drove young warriors keen to see lions to an area circled by tourist vans. The lions, lazing around unfazed by the yelling tourists, bolted like scalded cats as soon as they caught wind of the warriors' red ochre hair and heard the distinctive cadence of Maasai chatter.

You must keep a detailed record in your notebook, Parashino insisted as I struggled to keep up with his flood of information. You are a scientist, and scientists take notes to remember things. I told him I intended to learn as he did, through the eyes of a cow, not from a notebook.

Over the years Parashino and I sealed our growing friendship in the traditional Maasai way, by exchanging cattle and calling each other *pakiteng*, a public declaration of our bonding. I was aware of Parashino's ulterior motive—to win me over to the Maasai cause, banking on my research to counter conservationists lobbying for Amboseli to become a national park—but that motive became incidental as our relationship matured and our trust deepened, ultimately paying dividends for both of us when it came to Amboseli's future.

Herding Sotwa and Matingab became a bridge between our two worlds. I probed Parashino for his insights on Amboseli, and he pumped me about my research and the outside world edging onto his land and into his life. At first I only vaguely sensed the importance of grass conservation in the survival of Maasai livestock and the role a herder's social networks played as an insurance against bad luck and hard times. Out herding, I began to piece together the finely interwoven movements of livestock and wildlife through the seasons. From my aerial and ground counts over the years, I built up compelling evidence showing livestock and wildlife numbers oscillating in lockstep with the undulating rhythm of rainfall.

Rainfall governs the pulse of life in Amboseli. The first storms of the March long rains and the October short rains trigger the migrations. The erratic scattered storms entrain the herds of wildlife and livestock seeking out the greenest flush. Once the water pans dry and the short sweet grasses wither, wildlife, trailed by livestock, gravitate to the courser grasses of the Amboseli swamps to tide them through the dry season. For the Maasai, drought refuges like Amboseli, known as *olkeri,* are key to the survival of their herds. Only once the outlying grasses are depleted do the elders authorize access to the late-season pastures. Errant herders are fined and shamed if they break ranks.

I came to admire the resilience and adaptability of the zebu cattle. Selected for their resilience to harsh conditions, these small humped cows with dewlaps like an old man's chin fold, Asiatic eyes, and stumpy horns are as varied in color as domestic dogs and as docile as ponies. I learned more about the finely tuned Maasai feelers for every nuance of the savannas from a season out herding Sotwa and Matingab than I did in years of research.

Despite my tutoring, I was still puzzled that herders walked their cattle up to twenty-five kilometers to graze and watered them only every two or three days in the dry season, when zebra and wildebeest were doing just the opposite—moving as sparingly as possible. I could understand the advantage of the herders grazing their cattle beyond the foraging range of wildlife held back by their young, but surely a three-day round trek to and from water was overtaxing the strength of even the hardiest zebu cattle. Just as baffling was the remarkable ability of zebu cattle to produce milk long after breed cattle would have succumbed to drought.

Lost for an explanation, I teamed up with Virginia Finch, a young American environmental physiologist studying the response of cattle to droughts. To test the strain of long-distance treks and infrequent watering on cattle, Virginia restricted two herds to a half ration of food, then ran each on a treadmill. One group she walked the equivalent of eight kilometers and watered daily, the other she walked thirteen kilometers and watered on alternate days. Three months later Virginia was startled to find that animals walking twice as far and watering half as often had lost no more weight than the less stressed animals.

What could explain the baffling anomaly? Virginia found that animals walking longer distances cut their metabolic rate—their energy outlay—by close to half. The lower metabolism offset the costs of walking longer distances and less water consumption. A lower metabolism slows the passage of food through the gut and increases the digestion of rough forage, enabling the cow to extract more nutrition from poor pastures. The energy and water conservation of cattle increases their survival chances in droughts and speeds up their weight recovery during the rains.[3] Evidently the Maasai knew far more about survival and production in the arid lands than biologists like me, taught that warm-blooded animals maintain a constant metabolic rate. Dieters, struggling to lose weight only to find they must eat less and less after shedding the first few pounds, are up against the ghost of our own evolutionary adaptations to food shortages in the savannas.

Herders get the most out of their animals by separating them according to their feeding needs and strength. Children take care of the small calves and sick animals close to the settlement where pastures are fenced off in an *olopololi*, pastures conserved as late-season grazing in the dry season when the earth around the corral is trampled bare. Milking females are usually kept within a half day's walk of water to ensure a supply of milk for the homestead. In the harshest times warriors herd bulls and nonlactating cows up to two days' walk away to reach the most distant pastures.

The deeper I delved, the more ecological efficiency and conservation practices showed up in Maasai husbandry practices. Anthropologists refer to the Maasai as milk pastoralists—herders who live off the milk of their herds rather than slaughter animals for meat or commercial sale. Milk converts fodder into protein over twice as efficiently as meat and explains the high density of pastoralists in the African savannas—ten times greater than commercial beef ranches. Visiting American cattle ranchers are dumbfounded by the number of Maasai living on the land, and when I accompanied a group of Maasai to visit the cowboys in Arizona, they were baffled by the vast open rangelands devoid of people. The difference hinges on the economy of rancher versus pastoralist. For the rancher, the cow is a financial proposition. The fewer his ranch hands the higher his profit. For the pastoralist, the cow is his diet and currency—his sole means to negotiate for more wives and support his family. A big herd and a large family bring him not only prestige and influence but more hands to manage his herd. A big herd is also security against hard times through the reciprocal livestock exchanges and mutual support networks he can muster with many animals.

Reciprocity, as I learned, lies at the core of Maasai success in exploiting the savannas, surviving hard times, and staving off warfaring competitors. Just as I had exchanged cattle with Parashino and established a tight bond of friendship, so Maasai bonds are tied figuratively by a cow-calf umbilical cord. An exchange of sheep is the first bond of friendship and reciprocity, and an exchange of cattle is a yet closer tie, obligating partners to help each other out, come what may. Other forms of reciprocation range from lending cattle to someone in need and being paid back when the person's luck picks up to placing vulnerable animals in an associate's herd (*enkta-aroto*) for safekeeping. A herder may lend a milking cow (*ketaaro*) or breeding bull (*aitogaroo*) to a friend worse off than himself. Or in bad years, times when his herd is too small to feed his family, he may offer his own children to help tend a neighbor's herd. *Osotua,* the most altruistic act of all, is giving cattle to a hard-up person in your network with no expectation of any return.[4] If reciprocity is the glue that binds together the Maasai peoples, ecological savvy is the bedrock of their survival and adaptation in the savannas. Virginia and I found a particularly colorful adaptation boosting the productivity and resilience of their herds: energy coloration.[5] The Maasai have an elaborate nomenclature describing the varied colors and patterning of the cows: black, white, gray, red, brown, brindled, and multihued combinations. Reasoning that black cows absorb more heat and drink more water than white, Virginia asked me to check out her expectation that fewer white cows were dying during the drought. I first asked Parashino his opinion. No, he told me. Water isn't the problem. Our cattle die of starvation. White cows don't do as well as black when they are weak and the nights cold. You will find more white cows are dying than black.

He was right. Intrigued, Virginia and I surveyed cattle colors from the temperate southern highlands to the arid northern lowlands of Kenya. A consistent pattern emerged. The hotter and drier conditions of the lowlands favor the survival of light-colored cattle; the cooler and humid highlands favor the growth and survival of darker cattle. In the lowlands, light cattle reflect more heat than black, drink less water, and forage further in search of pasture. In the cooler highlands, black cattle act like lizards, absorbing heat in the early mornings, and so produce more milk and grow faster than white cows. The color adaptations to heat, cold, and water stress give pastoralists remarkable adaptability in raising herds across a wide range of climates from the baking Chalbi Desert to the cold highlands of Mount Kenya.

Social networks, ecological savvy, and livestock adaptations are the hallmark of pastoral survival, production, and resilience, but other conservation practices are also vital, including digging and protecting water holes, watering animals in hand-dug troughs to avoid wastage and pollution, drying and storing meat, supplementing milk with blood drawn from a cow's jugular in harsh times, conserving trees to ensure browse for their animals in the dry season, and pollarding branches to fence in and protect livestock from predators. Everything used is recycled back to the earth. When a settlement is abandoned, the thorn fencing, saplings framing the huts, and the dung accumulated over years decompose and fertilize the ground, creating islands of rich pasture.

Family survival and well-being are intimately bound to the welfare of livestock, the productivity of pastures, and the availability of water. Oddly, the Maasai have no word for conservation. Their lifeline—running from rain to grass to the productivity of their herds, the welfare for their family, and the health of the land—is summed up in one word: *erematere*. Erematere is an ethos, a way of life that blends Maasai husbandry practices, culture, and mindset.

Despite all that I learned in Amboseli, one big puzzle remained: how can such a plethora of wildlife and livestock coexist in the savannas? Surely the rational herder should get rid of wildlife to increase his own herds. Why, then, did the Maasai shun killing wildlife?

I found no answer to the paradox until I was out herding one day with Parashino in the depth of drought. We were tailing the herd, watching our cattle nuzzle over the sparse pasture. Fifty paces away zebra and wildebeest grazed, unfazed by the cattle bells and Parashino's piercing herding whistles. Puzzled by his indifference to wildlife vying with his cattle for pasture, I pumped Parashino for an answer.

Before the colonial government banned hunting in the 1920s, he told me, the Maasai survived the worst droughts by hunting wildlife, *in'gwesi*. They bought bows and arrows from the Sonjo on the Nguruman Escarpment and were taught how to hunt. Their favored prey were eland and buffalo, animals most like cattle. Wildlife was considered second cattle, animals the women once owned and let stray. The *laiboni*, the spiritual leaders, told herders to spare wild animals so that they would be available to stave off drought when the cattle died.

I was rocked by his revelation and deeply skeptical. Deferred gratification flies in the face of vested self-interest, at least according to the dictum that a bird in the hand is worth two in the bush. The deeper I delved, though, the more Parashino's explanation made sense of Maasai tolerance of wildlife. Time and again elders I queried repeated Parashino's story and lamented the loss of erematere knowledge among the younger generation. The narrative also matched accounts by early European explorers of the Maasai hunting and gathering when their herds died in droves during the severe droughts and rinderpest pandemics of the 1880s.

I was enthralled and buoyed by the notion of wildlife as second cattle. The expanded erematere ethos not only explained the enigma of coexistence but gave me a way of using Maasai conservation practices to win a place for wildlife without carving up the savannas into parks and ranches, severing the migrations. I wrote a couple of articles for science journals and conservation magazines arguing that plans for wildlife conservation should rest on the Maasai traditions of coexistence.

My induction to Maasai traditional practices paid off as Parashino had hoped. I gave a public lecture in 1969 at the National Museum in Nairobi laying out plans to sustain Amboseli's wildlife migrations by making the Maasai beneficiaries of wildlife tourism, ensuring them a cash income to complement their livestock economy. The proposal drew a lukewarm response from the conservation community but aroused the keen interest of development specialists at the Institute for Development Studies in the University of Nairobi. The institute had drawn in a radical group of political and social scientists, anthropologists, and economists bent on finding development paths suited to Africa. You need a strong economic case to win over the government, they advised me: Maasai benefits from tourism income alone won't win over the Treasury.

I was fortunate to link up with the Canadian economic advisor to Kenya's Ministry of Finance and Planning, Frank Mitchell. Fired up about social injustices and a keen birder, Frank jumped at the chance to visit Amboseli, flesh out a full economic justification for my proposal, and add to his bird list. His report quantified the gains for wildlife conservation, the Maasai, the tourism industry, and the government by expanding tourism use from the 250 square kilometers of the Amboseli Basin to the 8,000 square kilometers of the ecosystem.

Frank's analysis was compelling. The projected income from the expanded visitor capacity stood to triple revenues from wildlife enterprises. The Maasai would gain a monetary source of income at a time when they were beginning to move into the cash economy without having to sacrifice their lands, and until tourism generated enough income to cover their livestock losses to wildlife, they would be paid a guaranteed annual offset fee.

The economic analysis Frank Mitchell and I cobbled together hit the front page of the national press: "The Money Locked Up in Amboseli," read the

East African Standard headline on July 4, 1969. The article caught the eye of the government, but not in the way we intended. Mesmerized by the projected $8 million to $14 million in tourist revenues, in 1974 President Jomo Kenyatta declared, without warning, that a 320-square-kilometer area of Amboseli would become a national park. All hell broke loose. The Maasai, riled at losing their dry-season grazing lands, speared dozens of elephants, buffalo, lions, leopards, and hyenas to vent their anger. I was devastated at the land grab and feared the worst for wildlife and the Maasai.

By good fortune, the Amboseli plan Mitchell and I proposed won the support of the New York Zoological Society and the attention of the World Bank. With Frank Mitchell's inside track at Kenya's Ministry of Finance and Planning, the backing of a prestigious conservation organization, and the promise of a major loan from the World Bank, the Ministry of Tourism and Wildlife approved the plan.[6] By 1977, when Amboseli formally became a national park, I had enough standing with the government and the Maasai to broker an agreement assuring the community a share of the park's revenues and the opportunity to set up its own tourism ventures. A 390-square-kilometer national park would be set up by the government to protect the swamps and woodlands wildlife used in the dry season from agricultural encroachment. The remainder of the ecosystem would be owned collectively by the Maasai as group ranches where the traditional occupants were granted title to these lands. The group ranches would be paid an annual fee for conserving wildlife on their land and encouraged to set up their own tourism lodges and enterprises to complement their livestock economy and reduce the visitor pressure on the park.

A year later, at the opening of Amboseli's first school built by revenues from the park, the aging leader Mweyendet captured the mood of a Maasai gathering: Today you have shown us that wildlife can again become our second cattle as they were in days long ago, he noted. We know that in the droughts when our cows run dry, we can depend on the milk of our second cattle. The money we get from wildlife will build our schools and hospitals. The park will provide the jobs our children need.

Despite this victory in winning back the traditional value of wildlife for the Maasai and in sustaining the seasonal migration, a far bigger cloud hung over Amboseli and the East African savannas. The 1973 drought dragged on into a third and then a fourth year, the longest the Maasai elders could recall. By the formal launch of the Amboseli plan in 1977, the stench of thousands of dead cattle pervaded the plains. What future could there be for the Maasai, their culture, and wildlife when their way of life, which had conserved the savannas, was collapsing?

The answer came from another Maasai I befriended in the early 1970s, Kerenkol Ole Musa. Tall and regal, Kerenkol had the abilities, sensibilities, and leadership expected of a *laigwenani,* a warrior spokesman. Kerenkol saw the erematere link between land and pasture being severed by families growing larger than their herds could support. Reading the signs, Kerenkol was among the first Maasai in Amboseli to take up farming to offset his faltering herds.

We were seated outside Kerenkol's house at his Namelok farm in 1977, talking about our success in getting the government to pay the Maasai a portion of the national park revenues. You have helped us in Amboseli, he told me. You've gained us a share in tourism, brought the World Bank to meet us, and got the government to support a new policy of engaging communities in conservation. It's time for you to take a wife and have a family, he added. So why are you still out looking at wildlife? For Kerenkol conservation was about saving water and forage for his cows, making sure they survived and did well. His herds enabled him to afford two wives. For him money from tourism built the Maarba School and educated his children. He had reason to look after wildlife because it had become second cattle once again. But if for me wildlife wasn't about money, cattle, wives, and educating children, what did I get out of it?

Kerenkol's query rattled me, not because he doubted the community conservation programs, which were showing success in Amboseli by then, but because he didn't share my passion for lions and elephants. For Kerenkol, wild animals were second cattle and a daily threat. I had avoided showing my passion for wildlife by foregrounding economic justifications. That was a good enough reason for him, as it had been for his ancestors. All the same, I was troubled by Kerenkol's question. Why was I so passionate about saving the animals I once loved to hunt? In the back of my mind stirred another, larger question: if conservation underpinned Maasai survival, surely it must have played a role in humanity's success as a savanna species and rise to global dominance?[7]

2

Consuming Passions

In response to Kerenkol's question of why I was still out studying wildlife and not raising a large family, I had to think back to my childhood, tracing my faltering steps from hunter to wildlife conservationist. Those steps, I realized, tracked the changing views of wildlife from our hunter-gatherer ancestors to our modern sensibilities about nature.

My first passion for wildlife began with hunting in the woodlands of southern Tanzania. As a kid, life was all about the thrill and skill of tracking and stalking big game, sitting around the campfire at night listening to stories of big tuskers and famous hunters. Among them were famous snake man Ionides, renowned ivory hunter turned warden George Rushby, and the hermit naturalist Ronald de la Bere Barker. For all my admiration of white hunters, African hunters like the Hadzabe, who guided us on elephant hunts and lived by their wits and skill, stalking big game with bows and arrows, intrigued me far more.

My early days on the hunting trail were as cut and dried as firewood, all raw emotions with no thought of animal suffering or the future of big game. Then, aged ten, I had an encounter with an animal that upended my rifle-barrel view of animals. I had dropped behind my father's hunting party to take a leak when a sable antelope stepped out of a thicket a few yards away and stared down at me, still as a statue. Scared and mesmerized by the enormous coal-black animal with sweeping scimitar horns, I froze. Moments later the sable gave a nod of its head and melted into the thicket. In those fleeting seconds I sensed in the sable a fear and a curiosity like my own. That was the beginning of my development of a new sensibility for wild animals.

My emotions took another turn when I made my first kill at fourteen. The stalk and the kill went perfectly, and the warthog dropped like a stone. Moments later, posing for a photo, I felt more murderer than hunter. How could I have snuffed out the warthog's life for a momentary thrill? My hunting days were now over—I switched from gun to camera and developed a deep love and respect for wildlife.

My father's own hunting urge waned with his worries over the growing bushmeat trade in Mikumi in the 1950s, driven by the expanding urban population of Morogoro Township sixty-five kilometers up the road. His newfound ardor for catching poachers and creating a national park at Mikumi was infectious. I was soon caught up in his adventures as an honorary ranger in the Tanganyika Game Department, sharing his worries about poachers killing off wildlife even as I sensed a growing reaction to his conservation enthusiasm. Villagers were more inclined to welcome ivory poachers who killed elephants raiding their farms than wait in vain for the Game Department to drive them off.

In my progression from hunting to photography, I added yet another layer to my passion for wildlife: the joys of natural history. Watching animals rather than tracking big tuskers sparked my curiosity in animal migrations and the wealth of wildlife in the African savannas. My newfound passion scuttled my plans to become a game warden and landed me at Leicester University in Britain in 1964, studying for a degree in zoology. Soon my fascination with animal behavior and ecology added another layer to my life's passions: science.

After I had arrived back in East Africa in 1967, shortly after independence, I was dismayed to find how little had changed since the colonial days. National parks were still making enemies where they could least afford it, on their borders among displaced communities. Now, reflecting on Kerenkol's question, I had every reason to feel pleased that the Maasai in Amboseli once again regarded wildlife as their second cattle. So why was Kerenkol's question so unsettling? I realized with a jolt that I had buried my own self-interests in saving wildlife under layers of rationalization and justification for fear of appearing sentimental about animals and uncaring about people.

Once I admitted my own reasons for conserving wildlife, I no longer had to mask my sentiments behind a veil of rationalizations. I could talk freely to Kerenkol of my passion for wildlife, just as he did for cattle. Leveling with Kerenkol not only changed my relationship with him, it also alerted me to ways of conserving wildlife among people who didn't share my sentiments.

With age and experience, my hopes have matured. My wish today is for my children to enjoy a fulfilling life, regardless of whether wildlife is as meaningful to them as it is to me. At our home bordering Nairobi National Park, we are surrounded by wildlife. I take a break from writing these lines to walk the trails along the river. I pass a herd of grazing zebra and a couple of giraffes browsing on an acacia and then, dropping down to the river, come upon a troop of baboons mobbing a leopard making off with an infant. My freedom to live safe and secure with wildlife at my doorstep is not the choice of most Kenyans, though. Fifteen kilometers across the park, the sprawling metropolis of Nairobi is home to 3 million Kenyans who prefer the city to the bush.

From an animal's point of view, no one reason for conserving wildlife is superior to another. Reasons differ from person to person, place to place, and can change over time, as in my case. The reasons and means we use to conserve wildlife have grown in modern times from hunting regulations to national parks, zoos, gene banks, conservation easements, and dozens of other philosophies and tools. Each reason and each tool should add depth and breadth to conservation, and each should be judged by the same metric: does it work or not? Motive matters less to me than whether any justification—animal rights, welfare, sport, subsistence hunting, recreation, romanticism, spiritualism, deep ecology, or biodiversity conservation—succeeds or not. We should judge the outcome by how many animals, habitats, or wildlands our actions conserve, not how passionately and firmly we hold to our beliefs and methods. The more reasons to conserve and the richer our tool kit, the greater the chance of saving the wealth of nature in our pluralistic world.

You get a measure of whether conservation works or not by censusing animal numbers, tracking their movements, and recording their condition and stress hormones. Taking stock of an animal's response is a better measure than the lobbying power of advocacy groups that seldom confront the cost of living with wildlife in their backyard. Professing a love of animals is one thing, living with animals threatening your life, family, and livelihood quite another. Kerenkol's record of conserving wildlife as a drought failsafe is evidenced by the abundance of wildlife on his land, even if his motive is abhorrent to animal lovers opposed to any killing.

My pluralistic philosophy may seem shallow and even permissive to conservationists who hold strong and singular views, but a lifetime conserving wildlife among varied cultures and in diverse settings persuades me that my passion is not enough, however strongly held. Pit my love of animals against the agony of a mother's loss of a child to a hyena, the rage of a herder losing forty goats to a pack of wild dogs, or a farmer's loss of his season's crop to elephants, and my emotions wither before theirs. Out of experience and necessity, my conservation creed, like conservation itself, is multilayered. Dig deep into a wildlife conservationist's past and you will find a few thick emotional strata overlaid by layers of rationalization. Having excavated my own strata, I no longer feel contradictory in holding sentimental, intellectual, and moral feelings for conserving wildlife—any more than I do in enjoying a good movie that moves me to laugh, cry, think, and reflect on life lessons.

I would have accomplished little as a wildlife conservationist had I insisted that others share my passion or failed to have compassion for victims of animals. As a director of wildlife, how do I turn a blind eye to a rogue elephant killing a farmer protecting his crops and explain to his wife that it is against my philosophy to remove the animal and ensure she is compensated by the government? Passion counts, but without compassion for wildlife victims, it is inhumane.

I was first struck by the evolutionary novelty of humans conserving other species in the grove of fever trees where I first pitched my tent in 1967. For eighteen months I heard the sonorous *boop-boop* of the Verreaux's Eagle-Owl at night and the constant chatter of Hildebrandt's starlings during the day, accompanied by the monotonous, incessant *tap, tap, tap* of Cardinal and Nubian woodpeckers searching out grubs in the desiccated boughs. Years of research later, I was able to show that elephants fleeing poachers, trying to reach the safety of the park, were destroying bird-rich woodlands. The finding raised a passing thought I would later explore more fully: do any other species share our human sensibilities for nature and other creatures? Would an elephant share my thrill in sensing the sentience of the sable antelope or care a hoot about the eagle-owls and starlings it had put to flight? I doubted it and wondered how the seemingly perverse urge to save other species, including dangerous animals such as elephants, bears and tigers, arose.

William "Bill" Conway, then director of the New York Zoological Society (now the Wildlife Conservation Society), gave me reason to dig deeper into the anomaly. Bill, the most renowned and visionary zoo director in America, joined the society's small band of field biologists in 1977. In an offhand moment he asked me a question he had been brooding over: can you give me a good reason for conserving biological diversity?

I told him I would give his question some thought and get back to him. The more I thought about his question, though, the harder I found it. Protecting the large, charismatic elephants and lions of Amboseli using tourism revenues to win over the Maasai is one thing. Saving biodiversity— all species everywhere regardless of their value—is quite another. Mulling over Bill's question about saving biodiversity and Kerenkol's about my passion for wildlife, I realized how little thought I'd given to why we conserve other species at all. Forty years later, I am finally getting back to Bill with a few answers.

My search began with an admission. Having met considerable success in saving wildlife and recognition for the papers and books I'd published, I had good reason to call myself a conservationist. Yet time and again when I called myself a conservationist, I met outright hostility. The hostility came from conservation seen as pro-nature and anti-development by politicians, farmers, and businesspeople in America as well as Africa. Even among wildlife enthusiasts, calling myself a conservationist could be unsettling. Martha Honey, who directed the International Ecotourism Society for several years and wrote one of the finest books on ecotourism, had invited me to give a keynote address along with the Green Belt Movement's Wangari Maathai at a conference on the role of tourism in promoting conservation and community development held in Arusha, Tanzania, in 2008.[1] Over lunch Martha asked me how I would describe myself, given my promotion of local communities in wildlife conservation. A conservationist, I told her, thinking I would have a sympathetic ear.

Martha cringed. Conservation doesn't begin to describe what you do, she responded. It ignores the social and political causes you champion. I pressed her about her worries. She was taken aback by my description of myself as a "conservationist," she said, because conservation is often taken to mean conserving nature by keeping out local indigenous people who are viewed as the problem. Citing Mark Dowie's 2006 article "The Hidden Cost of Paradise," which excoriates conservationists for the harm national parks cause indigenous people, she was concerned that I would be typecast and misunderstood.[2]

Dowie was not the first to slam conservationists for being blind to human suffering. Raymond Bonner in *At the Hand of Man* (1993) lambasted Western nongovernment organizations (NGOs) for saving wildlife at the expense of African communities and failing to train African conservationists and scientists. Marc Chapin, writing in the World Watch Institute magazine in 2004, criticized the big international NGOs, the BINGOs, for ignoring the plight of local communities displaced in the name of wildlife.[3]

I had no wish to be typecast as an environmental fundamentalist or animal rights extremist. Out of necessity and compassion, even the most ardent wildlife conservationist must adapt. In setting up the Wildlife Planning Unit for the Kenya government in 1977, I ran up against politicians furious at me. Listen, development comes before conservation, I was told. You've got dozens of parks already. What about the landless people? Don't they come before wildlife? Later, as director of the Kenya Wildlife Service in the 1990s, I was attacked by animal rights groups for putting down marauding elephants killing people and destroying crops. "Use it or lose it" conservationists pushed me to lift Kenya's ban on sport hunting and ivory trading. Deep ecologists argued that I should take a biocentric rather than anthropocentric approach in Kenya's wildlife policies, meaning I should see the world from nature's perspective, not my own.

The narrow view of conservationists as nature-first activists is portrayed in the heroic images of Green Peace Zodiacs slamming up against whaling ships on the high seas and Dian Fossey battling poachers in the Virungas to save the mountain gorilla. In the process, the conservation professional who deals with energy, water and soil conservation, forestry, fisheries, and wildlife management, and the environmental engineer who curbs pollution and recycles waste have been lost in plain sight, replaced in the media by heroes like Jane Goodall, Joy Adamson, and Iain Douglas-Hamilton— people who save chimps, lions, and elephants. Others, like Wangari Maathai and Jacques Cousteau, have become icons for saving forests or oceans. Media figures like Paul Ehrlich and Al Gore work to save the planet from human overpopulation and global warming. Prominent in the public eye, all have become our modern conservation warriors fighting to save nature.

As a conservationist I welcome the attention the media lavish on saving charismatic species but am deeply troubled by the narrowing public image of conservation. Before recently amended, Wikipedia's definition of *conservation* defined a conservationist as "a person who advocates for the conservation of plants, animals and their habitats," a definition that entirely ignores the conservation of soils, water, farmlands, rangelands, forests, and fisheries that underlies the success of societies both ancient and modern.

I share the sentiments of nature conservationists and know how passion can spur people to extraordinary acts of dedication and courage. But we have done only half a job if we ignore those who lose out to wildlife or who have never had the chance to enjoy it as a tourist does. Joy Adamson's *Born Free,* published in 1960, topped the *New York Times* best-seller list for thirteen weeks. Adamson's heartwarming stories of raising the wild-born cub Elsa in northern Kenya made lions lovable in Europe and America. To my mind, for all her fame, Adamson deserves far more credit for her unheralded role in inspiring a generation of schoolchildren. In 1969 kids she funded on a nature safari to Samburu National Reserve were so thrilled to see wildlife for the first time that the group leader, Theuri Njoka, set up the Wildlife Clubs of Kenya. As patron of the clubs, I can attest to the millions of Kenyans who have become wildlife enthusiasts and crusaders through Adamson's dedication to wildlife education.

The future of our planet rests squarely with the 8 billion people whose small actions ultimately matter far more than a handful of ecowarriors and heroes of the planet, as important as they are. After all, it is consumers, you and me, who are the biggest polluters and destroyers of the environment because of our unrelenting hunt for the cheapest price in supermarkets and at the gas pumps, no questions asked, and because of the mounds of trash we pay garbage companies to collect and dump out of sight and out of mind. In the final analysis, small changes in the habits of billions can change the world.

I have in mind the individual who switches off lights and uses low-flush toilets to save energy and water and to cut costs, the person who installs double glazing to conserve heat, recycles trash, hikes, bikes, canoes, or fishes in the backcountry, visits a national park in summer, hunts deer in fall, fishes in clean running streams, and takes a trip to the Amazon rain forest, African savannas, or the Great Barrier Reef once in a lifetime. Then there are the big corporations, governments, and international institutions that are adopting the principles of corporate social responsibility and moving toward the green economy.

Conservation is done as a matter of course by the invisible legion of citizens, companies, and institutions whose tiny steps collectively add up to billions of small but vital actions. The legion includes those who pay the taxes that fund environmental monitoring, regulations, and enforcement, and those who work to conserve open spaces, nature, and wild places at home and around the world. The public has created a huge and growing demand for outdoor activities, whether visiting parks, skiing, hiking, fishing, or hunting. The scale can be gauged by product sales in the United States amounting to $646 billion annually. The ripple effects through the economy total $1.6 trillion annually, generating 12 million jobs and $80 billion a year in taxes. Each year 725 million people visit America's national parks and spend $20 billion in the surrounding communities. Put simply, the millions of Americans visiting the great outdoors create the biggest demand for nature and pay most of the costs of supporting it. The same can be said of the millions of international travelers whose expenditures of $2 trillion a year underwrite much of the developing world's conservation of cultural and nature heritage. Kenya's tourism accounts for 11 percent of its GDP, is the backbone of the country's wildlife industry, and attracts foreign investments in other industries.

These unwitting conservationists help to save lions and elephants every time they switch on a nature channel or mail a buck or two to the World Wildlife Fund or the Wildlife Conservation Society. Conservation is as much about the efficiency of resource use—treading lightly on the land and leaving a small ecological footprint—as it is about saving endangered animals and plants. In this sense the corporate world can play a role in conservation by cutting energy, water, and resource use as a matter of efficiency and cost savings, or as a show of corporate social responsibility for the common good. So, for example, the establishment of Environmental Protection Agency regulations in the 1970s saw lead emissions fall by 99 percent, sulphur dioxide by 81 percent, and nitrogen dioxide by 54 percent by 2015—despite a 57 percent growth in population, 246 percent growth in GDP, and 184 percent increase in vehicle travel miles, according to the Environmental Protection Agency's report *Our Nation's Air* published in 2016.[4] Increased efficiency in natural resource use, industry, agriculture, and waste recycling is an indirect form of conservation that lowers our ecological footprint, gives us a cleaner and healthier environment, and frees up more land for outdoor use and nature.

Reflecting on my own development and how my views have changed with the emergence of modern nation-states in Africa prompted my exploration into yet deeper questions: why are there so many different forms of conservation, where does the very notion of conservation come from, and how has it evolved to the point of saving other species and the health of our planet?[5]

3

The Conservation Paradox

September 2007. My wife Shirley and I have joined a small gathering of artists, musicians, writers, historians, sociologists, archivists, and philosophers in the seclusion of a Benedictine monastery on the island of San Giorgio Maggiore, looking across the Grand Canal to the Piazza San Marco in Venice. The meeting, "Coping with the Past," is hosted by the Cini Foundation and convened by Bruno Latour, an iconoclastic French colleague renowned for breaking down academic boundaries and synthesizing ideas from philosophy, sociology, anthropology, and biology.[1] Consider, he urges us, how we can faithfully conserve Venice's grand works of art, architecture, literature, and music against the ravages of time and changing cultural tastes. How do we preserve the past without betraying it?

The high point of the meeting is the restoration of one of Venice's grand masterpieces, *The Wedding at Cana,* a sweeping ten-by-seven-meter mural by sixteenth-century artist Paolo Veronese in which Christ miraculously turns water into wine. The work was commissioned by the Benedictine monastery to frame the back wall of the cavernous refectory of the basilica of San Giorgio Maggiore. Veronese so exquisitely fused the painting with the architecture that you feel you are seated at the end table among the wedding guests, mesmerized by Jesus performing his miracle. The painting was the pride of Venice until Napoleon's army sacked the city, slashed down *The Wedding at Cana,* and hung it in the Louvre in Paris across from another Italian masterpiece, Leonardo da Vinci's *Mona Lisa.*

Adam Lowe, a renowned English restoration artist, spent two years scanning *The Wedding at Cana* after the French government balked at repatriating the original to Venice. Using high-resolution digital photography, Lowe deftly stitched together each image on a computer and printed out a high-resolution facsimile to faithfully capture each brushstroke of the original. But which original? Should Adam replicate the Louvre painting, ravaged and darkened by time, or restore Veronese's original vibrant colors—a task made possible by research into the denaturation process of the dyes? Should he photoshop out the saber slash caused by one of Napoleon's looting soldiers or keep it? Which mural would Veronese think captured his artistry and purpose better, a color-restored facsimile housed in the hallowed sanctuary of the Benedictine monastery, or the darkened original hanging in the Louvre for public display, overshadowed by da Vinci's *Mona Lisa*?

Several art critics felt Lowe had betrayed the authenticity of the Veronese masterpiece, grouching that a facsimile is no substitute for the original, however blemished. The Venetian families we join at the gala to celebrate the restoration of *The Wedding at Cana* in the San Giorgio refectory on a sunny July evening disagree. They are in a rapturous mood, enthralled by Lowe's stunning replica. By public acclaim if not professional judgment, Lowe's *Wedding at Cana* is itself a miracle. It fuses the refectory tables laden with authentic Venetian cuisine and accompanying musicians dressed in period costumes playing restored sixteenth-century stringed instruments with the painting's tables piled high with food and wine to feed the wedding guests at Cana.

I am attending the "Coping with the Past" conference as an ecologist offering insights from the management of species and ecosystems to the conservation of Venice and its artifacts. The connection between the preservation of art and the conservation of wildlife seemed tenuous, but I accepted because I shared Bruno Latour's iconoclastic enthusiasm for breaking down disciplinary boundaries in the serenity of a Venetian monastery.

The arcane discussions in the cloistered retreat at San Giorgio seem remote from Maasai struggles to survive drought and my efforts to save wildlife. Yet the more we debate whether to allow cultural artifacts to age and deteriorate or to restore them to their original appearance and purpose, the more I'm struck by how freely the word *conservation* is used. Whether the issue is saving our own energy; economizing on car fuel, home heating, or water; saving whales, wilderness, and biodiversity; preserving cultural treasures like the city of Venice and *The Wedding at Cana;* conserving ancient manuscripts housed in the Papal Library; or Shirley's study of baboon troops in northern Kenya, the meaning of *conservation* is never in question. Yet for academics to use the word so liberally across the arts, humanities, and science without dickering with the terminology is as rare as stardust.

Standing on the banks of San Giorgio Maggiore with Shirley and Bruno, watching a Hilton-sized cruise ship ease through the narrows of the Grand Canal and dwarf the famous Piazza San Marco, I'm struck by a contradiction: the further we reach beyond our past and nature, the more we value them. The more we erase the diversity of cultures and the natural world, the more we do to conserve them. Venice, built on an archipelago of islands in the shallows of the Laguna Veneta, came to prominence as a mercantile and banking center in the twelfth century and grew into a city-state of resplendent architecture and refined art. Sidelined by long-haul jets, oil tankers, and container ships linking the new trade axes of Asia, Europe, and the Americas, the glory of Venice would have withered had it not been for a surge in international visitors.

Today, Venice is recognized as a World Heritage Site attracting twenty thousand tourists a day. An influx of wealthy foreigners buying up and restoring its crumbling houses has driven property prices beyond the reach of most Venetians. Rescued from neglect, Venice now risks being loved to death by hordes of visitors and becoming a living museum of restored monuments and artifacts lacking a Venetian soul.

The globalization of tourism has saved the physical charms of Venice at the expense of its ancient mercantile economy and culture, a culture vanishing because Venetians prefer the benefits of modernity over the hardships of living in a medieval city. According to purists, historical Venice has been betrayed. Like Veronese's *Wedding at Cana,* Venice has lost its authenticity, yet the city is valued by millions around the world today. Few Venetians in the city-state's heyday were rich enough to commission the works of a Veronese or a da Vinci. Today, even the poor living in the favelas of Rio or the slums of Mumbai can view Venice's grand masterpieces on TV or bring up an image of *The Wedding at Cana* on their smartphones.

Serengeti and Yellowstone national parks share much in common with Venice. Both ecosystems would have been lost to farms, ranches, mines, and settlement but for the change in sensibilities and an influx of tourists spawned by globalization. Both iconic parks are conserving the physical appearance of the wilds, yet they have lost the cultures and peoples who shaped them. Their descendants, who are joining the Venetians in their pursuit of modernity, are reshaping their homelands. National parks do conserve vignettes of the past, but their landscapes are laced by roads delivering hordes of visitors to luxury lodges offering all the modern conveniences of a Manhattan hotel. Like *The Wedding at Cana*, our national parks have been saved and transformed by their very popularity.

And that's not all. Four decades ago, we could hope to conserve the essence of Venice, Yellowstone, and Serengeti despite the annual influx of tourists. No longer. Greenhouse gases emitted by fossil fuels are expected to raise the Earth's temperature by two degrees centigrade or more and sea levels by a meter or more by 2100, according to a 2014 report of the International Panel on Climate Change.[2] The species of plants and animals making up the distinctive ecosystem assemblages of Yellowstone, Serengeti, and other parks are unlikely to move in lockstep with climate change, any more than they did in the past. More likely, the speed of human-induced global warming will produce novel climates and novel ecosystems.

Should Venice be left to drown, its treasures transposed elsewhere brick by brick, in the face of sea-level rise, or should the city be shored up against the rising waters by a tidal barrage? As with *The Wedding at Cana*, Venetians have decided on a local solution: the construction of an $8 billion seawall and sluice gates to regulate the tides and storm surges. The barrage may prove to be only a temporary reprieve if the Antarctic ice sheets collapse and raise sea levels three meters or more by the end of the century and thirty or more in the twenty-second century. What then? How many World Heritage Sites like Venice can we afford to save? How many Yellowstone and Serengeti parks can we protect as vignettes of the past in the face of climate change and other assaults if little is done to arrest our greenhouse gas emissions?

Like Venice, insular cultures and distinctive ecosystems are vanishing at a quickening pace. Species, our own included, evolved over millions of years, driven by geographic isolation and local adaptation. Globalization has erased differences in our ecologies and cultures so rapidly that conflation and homogenization rather than diversification are the new norms. The Tehuelche Indians Charles Darwin encountered in Patagonia on the voyage of the *Beagle* in 1832, like thousands of cultures, languages, and species, have vanished.[3] By the time Darwin set sail on his epic voyage, the great "Columbian Exchange" of species and domestic plants and animals, borne by sailing ships across the Atlantic and around the world, was well advanced.

In the last century, mechanized travel has sped up the exchange and jumbled up species and cultures. As with the construction of a Venetian tidal barrier, we can fence off parks, but walling them in will create ecological islands too small to sustain their former wealth of species. Preserving unique biotas and cultures of the world will take better knowledge and management skills than we have, and far more money than we can spare. The cost will become insurmountable if we don't soon contain the rise in global temperatures within tolerable levels.

The defense of Venice and the restoration of *The Wedding at Cana* epitomize the diversity, scale, and challenges of conservation. Conserving the past, whether art, architecture, music, language, culture, wildlife, or biodiversity, often depends on the very processes of globalization erasing them. We can't conserve the mounting knowledge, skills, and artifacts of humankind without the science and tools of modernity. We can't preserve anything immutably against the forces of entropy, decay, and transformation. But we can with concerted effort conserve the semblance and essence of the natural world and the necessities of our own survival and well-being, whether foods, health, energy, property, institutions, or cultural heritage.

Self-interest is the taproot of conservation. It takes escape from the daily struggle of hunting and gathering food to make a Venetian out of a !San hunter-gatherer—to have the time and means to create and then conserve the cultural richness of Venice and to embrace a larger vision to conserve all forms of life on Earth. This is not to say that traditional societies don't conserve species or habitats. Whether as clan totems or as pets, for their beauty or as insurance against times of scarcity, many cultures have conserved species they value. Many others still conserve large tracts of forest and lands they hold sacred. All the same, the !San, archetypal African hunter-gatherers who tail the migrating herds and shifting water sources of the arid Kalahari, would think it absurd to save a broken gourd or worn-out sandals and build a mansion to show them off, much less charge a fee to view them. Similarly, Venetians would think it absurd to store water in a gourd against drought when they can turn on a tap fed from a distant reservoir. The common ground is the !San hunter-gatherer and the Venetian conserving things that matter most to them.

The terrain becomes harder to recognize as conservation expands from its roots in necessity and expediency to include outdoor recreational space, the joy of the wilds, and our sense of well-being, and from its origins in community and neighborhood to world citizenry and planet Earth. Globalization has so stretched the length of our food and goods chain, and so disguised the link between our actions and their consequences, that the cultural sanctions and approbations our ancestors employed to reward good neighborly practices and punish bad ones are often lost. The Swahili fishermen of the Indian Ocean banned catches from their dugout canoes in the *milango,* the doors in the coral reefs, to ensure the tidal resupply of fish stocks on the coral reefs. Sanction breakers were banned from fishing, fined, or ostracized by their community. Today their descendants, lured by a cash income, use outboard motors and dynamite the reefs to stun and capture fish in order to compete with long-haul fishing fleets from China and South Korea.

At the same time, modern knowledge has so expanded our understanding of society and nature that we can now trace, track, and quantify our environmental footprint, including impacts unimaginably remote and inconsequential to our ancestors. Imagine the technology it took to detect the ozone hole over the Antarctic and trace the causes back to halogen gas in leaking refrigerators. Further, our modern sensibilities have stretched so far beyond utility that we now consider other species for their intrinsic value and strive to save them despite the costs and dangers.

The extermination of species by our ancestors, whether the mammoth and giant sloth in North America, the moa in New Zealand, the dodo in Mauritius, or bears and wolves in Europe, may seem irrational and tragic in our cushier age. Yet a moment's reflection shows that we are far more destructive in our pursuit of material goods and comforts than our ancestors were in their struggle to survive. We have razed forests and eradicated biodiversity to grow a handful of domesticated plants and animals, harnessed rivers to control floods and govern waters, and leveled landscapes to build roads and cities. The taming of the Wild West by hunters and homesteaders, aided by the U.S. Army, was often touted as America's proudest achievement in the nineteenth century. Conquering and banishing wilderness was the very hallmark of progress and advance of civilization, paving the way for yet more homesteading. Today nearly half the American West, from the Mississippi to California, is protected in national and state parks, forest reserves, wilderness areas, and wildlife refuges. Conservation easements are fast adding even more conservation lands across the United States.

The extraordinary change in our understanding of the world and in our sensibilities stands in stark contrast to the parochial interests and limited views of our great grandparents. A consideration of other species would have seemed bizarre and misplaced to the American pioneers who exterminated the wolf to protect their flocks and children. Spending millions of dollars to save other species still seems inhumane to half the world, mired in poverty and threatened by wild animals, pests, and parasites.

From the perspective of rural communities in the developing world, the government should spend its money combating bandits and rustlers; building schools, roads, and health clinics; creating markets for inhabitants' produce, rather than saving wild animals and evicting people from their homeland to create national parks. From the perspective of the urban poor, the government should focus on eliminating crime; eradicating poverty; and improving health care, housing, education, and social services, not on saving wildlife and wild places they will never have the means to visit.

Is it surprising, then, that marginalized societies still struggling to survive don't share the newfound sensibilities of many in the West for saving wildlife, biodiversity, and cultural artifacts? There is a deep divide between the poor, who see their prime responsibility as the self-preservation of family and tribe, and the Western conservationists, who argue that sacrifices must be made today to conserve nature for future generations. The dilemma confronting every conservationist is whether to put local people or the interests of the larger society and future generations first. As good neighbors we fail if we show no sympathy and offer no support for logging families in our town who end up on the dole because we prefer to save the spotted owl than to log the old grove forests owls depend on. As compassionate human beings we fail if we ignore the suffering of other communities evicted to make way for wildlife and national parks. As parents we fail if we don't put our own children above all else, including shooting an endangered species that threatens their lives, like elephants.

Weighing Kerenkol's conservation of his environment for the sake of his herd and family against my conservation of wildlife for the deep satisfaction it gave me challenged my very notion of conservation. Did my brand of conservation count more than his? And what of Bill Conway's mission to save all life-forms or the Venetians' desire to preserve their city and art? Such questions prompted me to explore the difference in their worlds and views: how conservation evolved from Kerenkol's brand, rooted in survival, to Bill's bid to conserve the diversity of life and the Venetians' their cultural heritage.

Whether it be Johannes Kepler's mapping of the planetary orbits, which Isaac Newton used to deduce his theory of gravity, or Carl Linnaeus's classification of species, which gave Darwin the scaffolding on which to construct the first tree of life, description is the starting point of understanding. Mapping out the broad dimensions of conservation sets the stage for answering the questions at the heart of my exploration. What is conservation? What role did conservation play in our emergence as a globally dominant species? And what conservation principles and practices might ensure our survival, safety, and well-being as we approach the limits of our planet's ability to sustain life?

Most texts on natural resources, the environment, and biodiversity treat conservation as a product of modernity. There are two problems with this view. First, it gives the impression that conservation is an invention of the West and depends on the West delivering the message to the rest. Like donor aid, the modernist view casts conservation as a north-south, rich-poor transfer of ideas, culture, knowhow, and funding. Second, it is simply wrong. As I learned during the droughts of the 1970s in Amboseli, conservation is latent and deeply embedded in the husbandry practices and survival skills of traditional societies.

Over the years, as I've traveled around the world taking stock of conservation traditions and practices, I've been intrigued by the tight link between livelihood and conservation in societies dependent on local resources for their survival, a link that becomes obscured and weak in modern market economies. The link is clear in the case of Bedouins of the

Arabian Peninsula, whose herd productivity and very survival depend on a few scattered oases. As the grasslands of the Sahara turned to desert some forty-five hundred years ago, the Bedouin tribes that adapted and thrived were those that worked collectively to dig an extensive network of underground channels, or *aflaj*, to distribute water to community members.

The connection between livelihoods and conservation is less apparent and more contested in contemporary California. Water rights divide farmers, homeowners, industrialists, developers, sports fishers, and white-water rafters. Each user stakes a claim to the common water resource. In 2016, during the prolonged withering drought of the 2010s, Governor Jerry Brown called on residents in towns and cities to cut consumption by 25 percent. The normally profligate urban residents came close to reaching the target. In contrast, farmers growing the water-hungry avocadoes, citrus, and walnuts that supply most of the nation used their economic muscle and prior water rights to remain exempt from rationing and continued to draw down a rapidly slumping ground-water table. If global warming projections of more frequent and persistent droughts in the American Southwest hold up, and they appear to be doing so, there will be no aflaj solution for California farmers: the water table will have been drawn down too far to sustain profitable fruit farming.

One of the great fissions in society since the birth of the modern environmental movement in the 1960s has been between advocates of unfettered economic growth and conservationists calling for crash programs to control population and development. Hidden in plain sight, conservation has become steadily more embedded in everyday modern life, whether in the form of urban planning for parks and recreation, regulation of waste disposal and recycling, control of pollution emissions, environmental impact assessments, fuel efficiency of vehicles, noise abatement of aircraft, or building codes to conserve or dissipate heat.

Efforts to separate conservation and development have been played up not only by conservationists but often even more vocally by industrialists and big businesses. Casting conservation in opposition to development abounds in contradictions. Good business practices embrace efficiency in the use of resources and production of goods and services to boost profits and shareholder dividends. Yet conservationists are often viewed as Luddites out to halt and reverse economic development. Boosting energy efficiency in our homes and industry to save money and reduce waste, wean us from imported oil, reduce the balance of payments, and improve national security—all are considered prudent and responsible behavior in the higher national interest. Why, then, is it imprudent for conservationists to want the same thing when it comes to shifting to renewable energy and to using our natural resources more efficiently and less injuriously?

The surprising thing about conservation is how essentially cultural it is, yet how ignored culture is in conservation. What we choose to conserve and how well we succeed varies among cultures and in time. Americans slaughtered the bison to the brink of extinction in the nineteenth century to drive out the Plains Indians and make way for white ranchers and farmers. Today half a million bison are back on the rangelands and the Blackfoot tribe, supported by conservation organizations, is poised to restore the first free-ranging plains bison in a century and a half. The bigger the goal and scale of conservation—and biodiversity is as big as it gets—the more people must work together. Conserving biodiversity is a cultural phenomenon unique to our species and our time, calling for global agreements and commitments that will be honored by future generations.

The connection between the preservation of art and the conservation of wildlife, which seemed so tenuous as we struggled to break down disciplinary boundaries in the serenity of a Venetian monastery, appears so obvious in retrospect: the diversity of conservation reflects both our dependence on and enjoyment of the natural world *and* the cultural artifacts and traditions of our own making.

Conservation in common usage speaks to its ancient roots in the survival and success of all societies no less than in saving biodiversity and sustaining our planet in an age of human domination. The ancient roots and universality of the word *conservation* are told in languages around the world. In Swahili, the most commonly used Bantu language in Africa, *linda* and *tunza* refer to protecting and guarding crops, livestock, and things of use and value. *Hifadthi* refers to conserving water, soils, trees, and the environment more generally. The *Oxford English Dictionary* traces *conservation* from its Latin root, "conservare," meaning to keep from degrading, decaying, and changing; to guard property. The same Latin derivative in French, Italian, Spanish, Portuguese, Dutch, and Romanian testifies to the handiness and hardiness of the word, and to the cross-cultural and interdisciplinary meaning of conservation that so intrigued me at the San Giorgio workshop.

Conservation has been key to every aspect of our success in rising to global dominance, and it will remain so if we are to create a future fit for our species and all others. Practicing and teaching conservation over the years, I've found that viewing it in evolutionary perspective, from its roots in survival to its expansion into environmental security, well-being, biodiversity, and planetary health, reveals the underlying unity of conservation despite its bewildering diversity.

Survival, environmental security, well-being, biodiversity, and planetary health: all are steps in the expansion of conservation down the ages, from our small-time primate origins to global domination. Such a viewpoint also helps in exploring the deep divides over conservation philosophies and practices.

There are, of course, other ways to classify conservation. Charles Redman, in his book *Human Impact on Ancient Environments* (1999), differentiates aspects of conservation by the role they play in the American mind.[4] Uppermost is how conservation can protect our health from polluted air, water, and hazardous wastes. Next comes the role conservation plays in aesthetics and outdoor recreation, and finally, its role in saving biodiversity. Redman's approach echoes an earlier classification by psychologist Abraham Maslow who, in his 1954 book *Motivation and Personality*, drew up a hierarchy of needs from the most basic sustaining us physiologically to the desire for security, love and belonging, esteem and, ultimately, self-fulfillment.[5] Redman's, Maslow's, and similar classifications are based largely on Western societies and worldviews. I take a broader view, reaching back to conservation rooted in the survival and competition of our species in the savannas and expanding in response to our changing role in nature and our own human nature in the global age.

I will distinguish between the various aspects of conservation—survival, environmental security, well-being, biodiversity, and planetary health—as a handy way to explain the differing views and deep divisions in conservation and how they might be reconciled. I start with the biggest debate in all conservation—the limits to growth. Are we prone to overpopulation and overexploitation to the point of collapse, or can we make a safe transition to a sustainable world?[6]

4

Limits to Growth: Hope or Despair?

The prospects of a population overshoot and environmental catastrophe have pitted prominent conservationists and economists against each other for decades. "Sometime between 1970 and 1985 the world will undergo vast famines—hundreds of millions of people are going to starve to death. Many will starve to death despite any crash programs we might embark upon now. And we are not embarking upon any crash program. These are the harsh realities we face." So claimed ecologist Paul Ehrlich in his highly influential *The Population Bomb* in 1968.[1] Julian Simon and Herman Kahn countered Ehrlich's apocalyptic outlook with *The Resourceful Earth,* published in 1984. "If present trends continue, the world will be less crowded (though more populated), less polluted, more stable ecologically, and less vulnerable to resource-supply disruption than the world we live in now."[2]

Is the world in imminent danger of mass starvation, as Ehrlich forecast, or does our future look brighter because of our technological ingenuity and the invisible hand of markets, as Simon and Kahn claimed? Or might we simply fall victim to our hedonism, selfishness, jingoism, and ethnic and religious intolerance? Will we lose half of all species to our rampant materialism, boom-and-bust economies, short-term planning, political myopia, runaway pollution, resource depletion, and global warming, or can we muster the foresight and wisdom to plan far enough ahead to ensure that our children enjoy a good quality of life rich in culture and biological diversity?

The conflicting views of optimists and pessimists in the Western world have fueled a long and rancorous debate over whether humanity can contain population and economic growth within the limits of available resources. The debate has divided poor nations, bent on catching up with the West, and industrialized nations, concerned about the impact of further population and economic growth on human welfare and the environment.

Birds go to great lengths to keep their chicks healthy by clearing their nests of parasites. The Dirty Nest Syndrome, as I shall call it, is a useful metaphor for our environmental impact as we transition from the ecosystem-scale communities of the Neolithic to the worldwide scale of the Anthropocene. At either scale we risk the well-being of our offspring unless we raise them in a clean and healthy world. Unlike in the contained world of our Neolithic ancestors, the lives of the remotest Mbuti communities of the Congo Basin are just as liable to disruption by our industrial age pollutants as school kids staying home to avoid Beijing's urban smog.

I vacillate between optimism and pessimism over whether people around the world can haul themselves out of poverty, better their lives, and restore their degraded homelands. In the 1960s, Machakos District, a subsistence farming area 60 kilometers east of Nairobi on the drive to Amboseli, was a basket case. Families of ten were crowded into makeshift mud huts. Every tree was hacked back or chopped down for firewood, every inch of land hoed and planted. Torrents of red soil washed off the slopes with each rain, and sinuous gullies knifed deep through the earth to bedrock, swelling the rivers with sediment and choking the coral reefs 150 kilometers downstream. As the drought of the 1970s deepened, I could see no way out for the Wakamba farmers any more than for the Maasai herders. It was then that I first read Ehrlich's *Population Bomb* and shared his prognosis of overpopulation, starvation, and collapse in Asia and Africa.

For all my fears, Machakos didn't collapse. On the contrary, farmers there have grown more prosperous and famines have waned. The barren, incised slopes are benched with terraces and erosion gullies are filling in. Fodder and fruit trees dot the landscape, and farmers are selling produce in Nairobi: doing well enough to build themselves a house, send their three or four kids to school, and buy a cell phone, TV, or even a motor bike, car, or truck. In 1994, Mary Tiffen, Michael Mortimore, and Francis Gichuki, in *More People, Less Erosion,* detailed the remarkable transformation of Machakos from the eroded hills of the 1940s to a productive market gardening community with five times as many people fifty years later.[3]

Machakos became a celebrated case of land restoration and social reformation. Forty years after publication of *The Population Bomb*, Paul Ehrlich wrote asking me if the evidence bore out the reported turnaround. I responded that the evidence is sound, the causes less certain. Stimulated perhaps by the opportunity created by roads and access to markets, the Machakos farmers began restoring their degraded lands. The president of Kenya, Daniel arap Moi, a farmer himself, rolled up his sleeves, took up a hoe, and launched a *harambee* (let's pull together) movement to build terraces and plant trees to arrest erosion, replenish soil, and conserve water.

Today Machakos, a bustling satellite town of Nairobi, has one of the most progressive county governments in Kenya and is emerging as an IT hub. Five fiber optic cables connecting Kenya to the rest of the world at lightning speed and a new IT city a short distance down the road from Machakos are propelling Kenya into the top ranks of the digital age.

Machakos tells a story of human ingenuity, cooperation, and national leadership restoring degraded lands and improving livelihoods. Danish agronomist Ester Boserup argued, contrary to the Malthusian theory that agricultural methods determine population density, that the reverse is true—population density drives up farm production. In *Conditions of Agricultural Growth: The Economics of Agrarian Change under Population Pressure* published in 1965, three years before *The Population Bomb*, Boserup argued that necessity is the mother of invention: "The power of ingenuity would always outmatch that of demand."[4] The two contrasting views fueled debates for years to come, though waning with technological advances like the Green Revolution and waxing with setbacks such as the world financial crisis of 2008, which sent world food prices soaring.

If the harambee spirit of pulling together gives Machakos and humanity grounds for hope, the same can't be said of a once cohesive community rupturing for lack of leadership at Shirley's baboon research site on the Laikipia Plateau two hundred kilometers north of Nairobi. Here in the Mukogodo Division, the population is still growing at a worrying rate; women bear seven children or more. The land is eroding deeper with each rain. Thickets of sisal-like *Sansevieria* and an invasive prickly pear from the American South, *Opuntia stricta,* are replacing trees cut down for corrals and fuelwood, and grasses are grazed to a nub by too many cattle, sheep, and goats no longer migrating with the seasons. Few families have enough animals to feed themselves.

After years raising money to build a boarding school and sponsor underprivileged kids, Shirley frets that poverty is creating habitual dependency on food aid. Famine relief is the community's lifeline in confronting shortening drought cycles. Most families facing growing poverty are hacking down trees to sell charcoal as well as river sand for cement and rocks for building construction. As our neighbor, veteran Maasai politician John Keen, puts it, the pastoral community is selling its natural wealth to buy poverty.

Add the external threats of war to the breakdown of traditional governance practices and population growth and the poverty trap worsens. I worry about the armed conflict in neighboring Somalia, where political anarchy, population growth, and poverty breed rustling, banditry, and famines. The reverberations are felt in Kenya, where former Somali warlords have sworn revenge on Kenya troops invading Somalia to drive the El Shabaab Islamic fundamentalists out of their strongholds. The defeated warlords have snuck into Kenya, recruited young Muslims on the coast, and carried out several attacks, including on the busy Westgate Mall in Nairobi and Mpeketoni on the coast. They have slain passengers on buses and quarry workers in Moyale, and gunned down over 120 students at Garissa University.

Pessimism thrives where once cohesive cultures have broken down, national governments are weak and corrupt, and tribalism is rife. Wildlife is as much a victim as people. Poaching gangs equipped with small arms bought for $30 from Somali warlords gunned down most of Kenya's elephants and rhinos. As director of the Kenya Wildlife Service in the 1990s, I worried over the fate of our crack ranger forces up against gangs of twenty or more poachers heavily armed with AK-47s. Each year several rangers lost their lives defending Kenya's wildlife.

In my conservation work I've tried to balance my Machakos optimism and Mukogodo pessimism. Why do we succeed in some cases and not others, and what does the past tell us about the future?

In 1798, thirty-two-year-old Reverend Thomas Malthus published a pamphlet that profoundly shaped our views of population and environment for the next two centuries, as world population exploded from 900 million to 7 billion. In *An Essay on the Principle of Population*, Malthus claimed that human populations tend to grow faster than agricultural production. "It is difficult to conceive any check to population which does not come under the description of some species of misery or vice," he wrote.[5] His conclusions laid the foundation for Charles Darwin's and Alfred Wallace's ideas on evolution and the struggle for existence.

Malthusian views fed the anxieties of the British aristocracy looking across the channel at the starving French peasants who had recently deposed King Louis XVI in a bloody revolution. In Britain poverty and disease cut life expectancy to a grinding forty years among all but the wealthy. In *The Wealth of Nations*, published in 1776, twenty-two years before *The Principle of Population*, Adam Smith expressed a similar view of the relation of food to population, which may well have influenced Malthus's dire outlook: "Every species of animals naturally multiplies in proportion to the means of their subsistence, and no species can ever multiply beyond it."[6]

In Adam Smith's view, the poor alone are held in check by subsistence: "In civilized society it is only among the inferior ranks of people that the scantiness of subsistence can set limits to the further multiplication of the human species, and it can do so in no other way than by destroying a great part of the children which their fruitful marriages produce." Although noting, as Malthus would, that farm production was closely tied to the natural fertility of the land and human labor, Smith tabulated a steady improvement in agriculture, manufacture, and commerce with the rise of capital investment, technology, and knowledge. The standard of living in the 1700s rose as the horse-driven plow replaced hand spades: "Not only grain has become cheaper, but many things, from which the industrious poor derive an agreeable and wholesome variety of food, have become a great deal cheaper."

Smith also recognized that poverty thwarted development. "No society can surely be flourishing and happy, of which the far greater portion of the members are poor and miserable. It is but equity, besides, that they who feed, cloath, and lodge the body of the people, should have such a share of the produce of their own labour as to be themselves tolerably well fed, cloathed and lodged." His warning would resurface in discontents over global inequality expressed two centuries later.

Neither Smith nor Malthus anticipated that the industrial revolution then getting under way would transform Britain and the world. Transportation and commerce, fueled by the combustion engine and wood and coal power, were about to launch a technological society no longer tightly tethered to the natural productivity of land or to the muscle power of humans and beasts of burden. Industrial fertilizers, insecticides, tractors, harvesters, and high-yielding crops boosted farm production in the coming two centuries, and medical advances cut disease losses and raised child survival rates.

By the late twentieth century, Europe's population had risen fivefold and the average life span twofold since Malthus's time. Per capita income had increased a staggering fiftyfold, and yet fertility rates in many countries were falling below long-term replacement rates. Most astonishing of all, Europe's fears of famine had given way to a surfeit of food. Death from overeating and overindulgence far outstripped mortality from contagious diseases. Human ingenuity had triumphed over the Malthusian shackle.

Or had it?

Malthus was wrong about the Western world, but his Hobbesian view of the short and brutish life of peasants was rekindled by scientists as populations surged in post–World War II Asia and Africa. In Kenya, for example, growth rates rocketed to 4.1 percent in the 1970s, a rate three times faster than Europe's at the height of the industrial revolution. The needs of the growing population far exceeded the customary farming output. Crop yields fell and erosion spread. The drought and famine that ruptured Kerenkol's traditional way of life and severed the erematere link between herds, family, and the land spread across from Kenya into Ethiopia and around the Sahel, killing tens of thousands and launching the Band Aid campaign, the largest famine relief effort in history.

Ehrlich's apocalyptic scenario of impending mass starvation was soon reinforced by the influential Club of Rome's 1974 computer model entitled *The Limits to Growth*, which projected population collapse and mass starvation in the developing world by the year 2000.[7] To Malthusian collapse, the Club of Rome added predictions of a steeply rising death toll from pollution and carcinogens. In 1962 Rachel Carson's *Silent Spring* had already alerted the public to the toxic fallout from insecticides, pesticides, vehicle emissions, industrial effluent, and heavy metals leaching into the soils, skies, and seas.[8] Barry Commoner's *The Closing Circle* in 1968 issued dire warnings of an overpopulated world running out of resources and industrialization poisoning the planet.[9]

The bald eagle, symbol of America and victim of eggshell thinning caused by food chain concentrations of DDT, was prima facie evidence of looming catastrophe.

Malthus's poverty trap proved wrong for Europe, and Ehrlich's wrong for Asia. Starting in the 1950s, new breeds of dwarf wheat and rice, coupled with increased applications of fertilizer and water, spawned the Green Revolution in the late 1960s, boosting yields and improving food security in India and southern Asia generally. Within three decades Asia bustled through a stunning demographic transition from high fertility and death rates to sharply reduced levels, boosting child survival and longevity and fast-tracking a transition from peasant farming to industrial economies that had taken Europe two centuries. Meanwhile, in China, the most populous nation on Earth, some 45 million people died of starvation in the late 1950s and early 1960s as a result of Mao Zedong's Great Leap Forward—his communist manifesto to boost agricultural production through collective farming. Forty years later China had leapt forward to become the world's second-largest economy in response to the Green Revolution and a market economy. World fertility rates in the second half of the twentieth century fell even faster than population growth rates rose in the first half, driven by contraceptive advances, family planning, China's one-child policy and, above all, the rising wealth, political liberalization, education, and growing role of women in the developing nations, leading to smaller family sizes and a better standard of living.

A measure of how rapidly the global outlook has changed can be gauged by forecasts made in the 1960s of a population of 30 billion by the end of the twenty-first century. Today, the Food and Agricultural Organization of the United Nations projects that the world's population will level out at between 9 and 12 billion by 2100. Pessimistic forecasts of rapid population growth still prevail in Africa despite the demographic transition well under way in most countries on the continent, including Rwanda, once seen as the basket case of the continent. Birth rates, still pegged at 3.5 percent in much of Africa, will see the continental population grow two- to threefold before settling to Europe's replacement rate at less than 1 percent. In the meantime, Africa's population will swell to 2.2 billion by 2050, and virtually every country will face Ehrlich's worst fears: how to produce enough food and find enough jobs for the massive surge in young people to stave off economic and social disasters.

How have conflicting scenarios of optimism and pessimism played out?

Although a quarter of the world is still locked in poverty, and global warming has climbed to the top of the international agenda, the unbridled optimism of Simon and Kahn has been rekindled by Matt Ridley's "Rational Optimist" (2012).[10] We are clever enough to keep on bettering our lives and inventing technical fixes to overcome shortages, clean up our mess, and fix our missteps, Ridley claims. A rising economy and new technologies will make it cheaper and more efficient to lower greenhouse gases later, rather than incur a heavy cost to the economy now, he argues.

In a similar vein, Danish political scientist Bjørn Lomborg, responding to the continued skepticism of catastrophists, amassed reams of statistics in his 2001 *The Skeptical Environmentalist* to show that human development indicators, ranging from levels of literacy and health to income and the quality of air and water, are improving with economic development.[11] Lomborg raised the hackles of the world's preeminent conservation biologists and promoted a withering response from dozens of scientists and environmentalists.

The ensuing exchange was confused, in my view, by the opposing sides arguing apples and oranges. Lomborg was right in saying that, on balance, development indicators are rising, but wrong in denying a biodiversity crisis. Conservation biologists were wrong in denying the improving human indicators, but right in denouncing Lomborg's embellished picture of biodiversity.

Lomborg's and Ridley's optimism over the prospects for biodiversity is rooted in the supposed feedback between economic growth and development expressed in the Kuznets curve. Simon Kuznets, renowned for co-developing the standardized measure of gross national product in the interwar years, claimed that increased income disparity emerges in the early stages of economic growth, but beyond a certain income threshold it begins to narrow.[12] A similar link was claimed between wealth and environmental health.[13] The rapid growth of U.S. agriculture and industry in the late nineteenth and early twentieth centuries did cause extensive destruction of forests and wildlife and pollution of the air, soil, and water. The burgeoning American economy did lift millions out of poverty in the 1950s, even as it left others languishing in poverty. And a wealthier, healthier America did launch an environmental movement in the 1960s, leading to legislation and executive action to reduce pollution emissions and set aside new lands for wildlife. Whether the link is causal is another matter. A Kuznets-style argument has been used by free-market advocates to claim that the invisible hand will lift the standards of all, clean up our environmental mess, and improve the lot of other species as a matter of course.[14]

Even if wealthier populations do begin cleaning up their own backyards, the laissez-faire response is far too slow for the Kuznets curve to reverse the global damage being done in the twenty-first century. The developing nations are ten times more populous and growing economically three times faster than the industrializing West at a similar standard of living. Beijing residents wearing smog masks and buying home air filters to avoid sucking in lung-clogging urban smog want government action now. Add to the urban respiratory threat the 16 percent of China's agricultural soils poisoned by insecticides and pesticides at a cost of $350 billion a year, and Kuznets's deferred cleanup is out of the question, even if the Chinese government is taking steps to curb urban smog and toxic soils in its five-year plan.

Disagree though we may with the Chinese government's draconian one-child policy, how much larger would the population be without it, how delayed the demographic and economic transition, and how much worse the poverty and pollution? Had Rachel Carson not raised public awareness, prompting government action, how polluted would America's rivers and the global atmosphere be, and how many tens of thousands more people would have died of respiratory diseases? How many more Three Mile Islands would it have taken for the nuclear energy industry to curb the risk of radiation leaks when the bottom line for the industry is shareholder profits, not public safety?

Advocates of laissez-faire solutions to our environmental negligence forget that the primary role of government is the safety and security of its citizens, and that our Dirty Nest recklessness is killing far more people than wars in our global age, to say nothing of the steeply rising medical costs.

Many of these issues surfaced three decades ago when I was turning my attention to the larger global threats to wildlife, the world's shrinking wild places, and global warming. I broached the idea of holding an international conference on the challenges facing conservation when Bill Conway asked me to direct the international programs of the Wildlife Conservation Society. Hosted by Rockefeller University in fall 1986, the meeting brought together a stellar assemblage of biologists, wildlife managers, planners, economists, development specialists, donor agencies, the World Bank, philosophers, and the media to size up the challenges of globalization in the twenty-first century.[15] Looking a few decades ahead, what new and unexpected threats and opportunities could we imagine? How could we build conservation into development plans, rather than rely on the premise of a Kuznets-style deferred mop-up after the damage is done?

In his opening address, renowned Harvard biologist E. O. Wilson quoted Fairfield Osborn's *Our Plundered Planet*, published in 1948: "The tide of the earth's population is rising, the reservoir of the earth's living resources is falling."[16] The pace of population growth and environmental destruction had continued with no visible pause since then, Wilson lamented.

In my own presentation I stressed that the best hope for us and all other species lies in the improvement of human welfare. If our outlook dims, we will cease caring about other species. If our lot improves, we will have more time, resources, and compassion for nature. The sooner we put wildlife on the agenda for the twenty-first century, the greater the chances it will find a lasting place in our world. Tarzie Vittachi, based on his experience in the UN International Children's Emergency Fund immunization campaigns, put the challenge in pithier language: the problem of excessive population growth will be solved not in the uterus but in the human mind. His lesson rang true in every field of development. Without people's willing engagement, sustained social change is a Sisyphean task. Changing our views on nature calls for a revolution in human attitudes, including conservation as a part of development rather than as an afterthought.

The divide between conservation and development began to narrow shortly after the meeting when the Brundtland Commission, convened by the United Nations, submitted a report on the principles of "sustainable development" to the World Commission on Environment and Development in 1987.[17] Despite the adoption of the principles linking development to environmental health, biologists' fears over global limits were rekindled by publication of Jared Diamond's *Collapse* in 2005. Taking up where his 1997 best-selling *Guns, Germs, and Steel* left off, Diamond asked why some societies rose to dominance and others failed.[18] He identified five main factors that may determine our fate: environmental damage, climate change, hostile neighbors, lack of friendly trading partners, and how a society responds to environmental problems. The first four are not always significant in explaining collapse, Diamond noted. A society's failure to respond to environmental threats always is.

Diamond drew on case studies around the world to test his "Big Five" theory. Among failed societies he included Easter and Henderson Islanders, Pitcairn Islanders, the Anasazi, the Maya, Vikings in Iceland and Vinland, and the Norse in Greenland. Next, Diamond contrasted these historical failures with notable successes—the New Guinea Highlanders, the Tikopians, and the Tokugawa Japanese Shogunate. Some ingredients of success were good land fertility, relative lack of social stratification and inequality, adaptability based on bottom-up solutions, and adaptive responses to climate change. Most failures involved inflexible lifestyles and a strong social dominance hierarchy. Diamond next looked at Big Business and the environment, drew up another list of factors contributing to their failure or success, and inferred the same factors would apply to the fate of cities like Los Angeles and the global environment. On balance, Diamond concluded, we face a dire future and the possibility of collapse in our hyper-populated world—unless we heed the characteristics of past success stories.

Archeologists question Diamond's views of collapse, arguing that we need more rigorous data to distinguish collapse from migration and adaptation to environmental disasters. Population movements often bridge the gaps between supposed collapsed societies and later societies, many of which are still in existence, they point out.[19] Whatever the answer, *Collapse* did not have the impact of *The Population Bomb* or *The Limits of Growth*, both because fears of overpopulation were receding by the 2000s, and because Diamond himself concluded that collapse, though very possible, was not inevitable: it all depends on how societies manage the environmental crises they confront.

Reading Diamond's *Collapse* reminded me of the hopelessness I felt about the disaster the Maasai faced in the drought of the 1970s and my surprise at their resilience, rooted in strong social networks and reciprocity. Unlike Ehrlich in *The Population Bomb*, Diamond recognizes the importance of social networks in environmental responses, though I believe he underplays their central significance.

A case in point is Rwanda, which Diamond cited as a worst-case scenario— a modern Malthusian disaster of overpopulation, land crunch, and resource shortage that triggered the genocide of eight hundred thousand—mostly Tutsi—people in 1994. Two decades later, Rwanda is one of the fastest developing countries in Africa due to a mediated reconciliation between the warring Tutsi and Hutu tribes, strong political leadership, the rule of law, economic liberalization, and economic incentives that have spurred initiative, innovation, and investment. Similar tribal clashes set back Uganda and Congo, countries with abundant land and resources. Uganda has since recovered for much the same reasons as Rwanda. Congo across the border is mired in strife and stagnation for lack of social cohesion, cooperation, and government leadership.

In weighing historical evidence and analogies about whether we shall overshoot and collapse or whether Boserup's power of ingenuity will prevail, let's consider our species' rise to global dominance in evolutionary perspective. Using the yardstick of the growth and spread of populations during the out-of-Africa diaspora some seventy thousand years ago, humans invaded every landmass and in the course of thirty-five hundred generations have become a geographically ubiquitous species. A population graph would show slow growth through the mid-Pleistocene until some fifty thousand years ago, a steady increase until the dawn of the Neolithic eleven thousand years ago, then a rapid rise with plant and animal domestication, and finally an exponential growth with industrialization, from a little over 1 billion to nearly 8 billion in the early twenty-first century.

The forces accelerating population growth from the Pleistocene to the industrial age can be gauged from the population densities associated with major shifts in means of livelihood. Hunter-gatherer densities average 0.1 person per square kilometer, traditional pastoralism 10 to 20, short-fallow farming 16 to 64, annual crop farming 64 to 256, multi-crop farming in the thousands, and today's megacities in the tens of thousands.[20]

When we break the data down by geographic region, yawning disparities emerge. Most regions show a steadily upward trend in population, with episodic downturns due to wars, disease, and catastrophes, such as the great tsunami that devastated Crete and the volcanic eruption that buried Pompeii during the time of the Roman Empire. Societies such as the Mayans and Anasazi perhaps broke down due to overuse of resources and climate change, as Diamond proposes. Whether the populations collapsed along with civilizations is an open question, though. Anthropologists tend toward the view that populations dispersed or were assimilated by neighbors rather than collapsed.[21] The weight of evidence since modern humans left Africa points to a sustained and accelerating growth, and the historical evidence in the Neolithic suggests localized declines but not large-scale collapse.

The problem in using the past as indicator of our future comes of treating small isolated communities and economies as analogs of large, open societies in the global age. As Diamond illustrates, small isolated and insular communities like the Easter Islanders and strongly hierarchical and despotic societies like the Mayans are vulnerable to breakdown in the wake of climate anomalies, natural disasters, and political turmoil because they lack the space to escape local disasters and the interchange with other societies that fosters cultural innovation and learning. On the flip side, biogeography and paleoecology show that wide-ranging generalist species and adaptable feeders like baboons are less susceptible to extinction than narrow-ranged specialists like the endemic Sokoke scops owl, a diminutive raptor confined to a narrow band of forest along the northern Kenya coast.

A better indication of how we might adapt to global changes comes from continental-scale responses to climate change and the importance of demographic mobility. During the African Humid Period, twelve thousand to five thousand years ago, archeological evidence and rock art in the Tassili and Ahaggar Mountains show cattle-herding cultures occupying a savanna landscape stretching across the present Sahara Desert.[22] Within a few hundred years the rain-bearing winds shifted, turning savanna to desert. Some communities adapted to oasis farming and camel keeping; others migrated to the Nile Basin. Yet others moved south into West and Eastern Africa, spawning a great diversity of farming and herding cultures.

An evolutionary and historical perspective on human growth and cultural adaptation highlights two points. First, our success did not depend on megamammals or biological diversity. Our species succeeded by clearing natural habitats, replacing wild with manageable domestic plants and animals, simplifying the food web, and boosting production with fertilizers and insecticides. Second, we must distinguish between different levels of conservation in judging success, whether for survival, environmental security, well-being, biodiversity, or planetary health. All living societies have, by definition, succeeded in the survival game. Whether we succeed in sustaining development and planetary health in the coming century is open to question.

I see the prospects of overshoot and collapse in the future as less likely than they have been in our past. We are no longer blind to sullying our nest and darkening our future. We have better tools to monitor our impact and project ahead. We have already established international and global institutions such as the United Nations to forge agreements on how to combat our impact on the global commons. The time has come to move beyond Malthusian fears and probe deeper into human nature to determine how we can make a safe transition from local ecosystems to global limits. We do need to admit the possibility of overshoot and collapse as a worst-case scenario if we behave like other species, and yet we must accept along with social scientists that our unique capacity as a species to foresee disasters and take aversive action can avert collapse.[23]

5

Lessons from Disasters

If in Darwinian terms we are so successful as a species, why worry when collapse is so rare and largely confined to geographically restricted, closed, and inflexible societies of the past rather than the ubiquitous open-networked societies of the global age?

Doomsday warnings hit a raw nerve and feed our deepest fears. I don't rule out the possibility of collapse from overpopulation and starvation, or from major economic and social disruptions caused by our environmental impact. There are simply more germane and existential threats to our well-being, whether wars, poverty, inequality, authoritarian ideologies, ethnic intolerance, corruption, greed, and heedless development. Unless we fix these economic, political, and social flaws, we will keep on polluting our environment, trashing biodiversity, and inviting collapse. If we fail to engage the cooperation and ingenuity that gave us the ecological edge over other species and distinguished successful from vanished societies, we shall fail to solve today's far more challenging global threats.

Collapse speaks to our fear of the raw ungovernable power of nature's cataclysmic forces told in our myths, legends, and religious texts. Natural disasters were wholly unnatural to our ancestors. Volcanic eruptions, earthquakes, hurricanes, droughts, floods, and tsunamis were acts of the gods to the unenlightened mind. Tempestuous forces were the gods at war, hurtling thunderbolts and lightning at each other across the sky and punishing humans for their irreverence and sins. Natural disasters augured collapse—not of populations but of something far worse; they signaled a descent into social anarchy and rapine.

The fear of decline and collapse is a recurring historical and literary narrative. In the Bible Matthew warns that when the angels separate the just from the evil at the end of the world, there shall be wailing and gnashing of teeth. It is worth remembering that Thomas Malthus was a preacher. Collapse and decline on a grand scale haunt nature lovers, biologists, and conservationists no less than theologians. Gregg Easterbrook argues in *The Progress Paradox: How Life Gets Better While People Feel Worse* that the declinist view is endemic in us and probably stems from our fears of loss and uncertainty.[1] Arthur Herman in *The Idea of Decline in Western History* traces the roots of our modern apocalyptic views to eighteenth-century French philosopher Jean-Jacques Rousseau, and on to the pessimism of Oswald Spengler, Arnold Toynbee, and Paul Kennedy.[2]

I learned more about nature, disasters, and human folly in the bush country around our school at Kongwa in central Tanzania than I ever did in class. We were playing on the games pitch one morning in the early 1950s when a brown cloud rolled over the horizon and enveloped Kongwa in a whirling melee of locusts. The three-inch-long biological defoliants known as desert locusts, *Schistocerca gregaria,* settled on every living blade of grass, corn, millet, and sorghum and stripped them bare. The plague began its 4,700-kilometer migration in the Arabian sands after sporadic storms and meandered south over several generations to reach East Africa in 1955. In the course of days, the locusts destroyed the livestock pastures and crops of the Wagogo farmers, bringing famine to central Tanzania.

Locusts swarmed in the dry thorn county of Tanzania every decade or two. The Wagogo are herders and farmers who originated in the Unyamwezi and Uhehe tribes in the nineteenth century. Resembling the Maasai in dress, the Wagogo take earlobe stretching to extraordinary lengths, literally. Using large wooden plugs that give them their Swahili name, People of the Log, the Wagogo stretch their earlobes down to their shoulders. Outwardly traditional in the 1950s, they were attuned to droughts, floods, dust storms, locusts, army worms, rinderpest, smallpox, and periodic cattle raids by the Maasai to the north. Disasters were a fact of life. Bumper years were a cause for celebration but, always with an eye to hard times, the Wagogo diversified their food sources, stored cereal crops in granaries, kept meat on the hoof and, like the Maasai, fell back on hunting and gathering in extreme droughts.

Wagogo knowledge and knowhow for surviving the hazards in Kongwa were ignored by the British government when it launched a grandiose peanut scheme across the tangled thorn country in the late 1940s, aiming to alleviate a postwar shortage of vegetable oil. On the advice of expert agronomists, the colonial government cleared tens of thousands of acres of bush using giant dozers hauling ten-ton metal balls, intent on scaling up the Wagogo desultory peanut harvest to industrial production.

Within two years the rains failed. The brick-red earth baked as hard as asphalt. Disc plows snapped and giant Caterpillar tractors clogged with dirt. High winds lifted the topsoil and whipped up roiling dust storms that enveloped the sprawling Kongwa settlement, sending galvanized roofs clattering across the barren flats like tumbleweed. Defeated by the African elements after the Allied powers vanquished the Germans and Japanese in the Second World War, the British Empire abandoned the ill-fated scheme and turned the town of clapboard and mud buildings into a boarding school. Within a few years the bush grew over the tractors, abandoned in the hasty exodus.

The Wagogo carried on farming and herding, muttering about the British government's folly in planting nothing but peanuts in the hopes of good rains. We only plant peanuts in good years, along with maize and melons, the Wagogo told us kids bartering corn to cook over our wood-fired water heaters. In bad years we depend on sorghum, finger millet, and goats, they recounted.

Like those of farmers across Africa at the time, Wagogo crop yields were paltry by modern farming standards, yet what they lacked in production they made up for in resilience to the erratic rains and hard-baked soils that defeated the imperial government. The Wagogo put the failure of the groundnut scheme down to human hubris and ignorance—not an act of God. For the traditional farmers the groundnut scheme was a passing moment; for the British government it was a monumental failure that squandered $100 million ($1 billion in today's dollars) in taxpayers' money and contributed to the defeat of Clement Attlee's government in the 1951 election.

Historians have documented the impact of spectacular natural disasters since the dawn of civilization. The tsunamis that swept away the Minoan civilization of Crete in 1450 BCE, the pyroclastic eruption of Vesuvius that buried Pompeii in 79 BCE, and the droughts that contributed to the collapse of Mayan cities in Central America between the eighth and ninth centuries are catastrophic natural disasters that have changed the course of civilizations.

Anthropologists and agricultural researchers have long recognized the risk aversion of traditional societies. The best insurance against natural disasters among hunter-gatherers, herders, and farmers is a strong social network.[3] You hunted, herded, and farmed with family and friends not only to grow more crops and maintain larger herds, but also because it was safer. More eyes and ears warded off wild animals and predatory neighbors, and assured help in the event of an accident.

The first rule of evolution is survival. Risk takers die young—the cautious are more likely to live, reproduce, and raise their young successfully. A poster in the flying school where I earned my wings showed a picture of a biplane crumpled in the branches of a solitary tree in an open field with the caption "There are old pilots and bold pilots, but no old bold pilots!"

When I visit New York, I'm astonished by pedestrians tuned in to a headset without a thought of danger. They have implicit trust in the social conventions and regulations enforced by fines that make sidewalks safe for pedestrians, roads safe for vehicles, and automated lighting symbols of pedestrian-safe crossing points. Then there are the lone hikers wandering into the Arizona desert or yachters off San Diego heading out to sea, relying on rescue services being a call away if they break a leg or capsize.

In our modern world, the seven plagues of Egypt forcing the pharaoh to free the Israelites from bondage are receding. The threat of locust, frog, fly, louse, and livestock plagues are vanishingly small today. Smallpox and rinderpest have been eradicated worldwide—with not a whimper from conservationists. The Rocky Mountain locust (*Melanoplus spretus*) that ravaged the crops of pioneer American farmers in the Midwest has been exterminated. Frogs, far from being a plague, are the most threatened of all vertebrates due to habitat destruction and a pandemic fungal disease.

The absolute death toll from disasters in our highly populated world is mind numbing compared to biblical times. The Aceh tsunami of 2004 killed 230,000, the Kashmir earthquake of 2005 some 100,000, and the Tohoku Fukushima tsunami of 2011 nearly 16,000. The growth and spread of populations are putting more people in the path of natural upheavals than ever before: between 1980 and 2014 the number of disasters caused by natural events rose from four hundred to nearly a thousand.[4] The death toll from starvation, on the other hand, has fallen steeply from the Sahelian and Ethiopian droughts that killed millions in the 1980s. International relief agencies prevented the 2010 drought from repeating the disasters of the 1980s, and today more people in the Horn of Africa die in civil disruptions and banditry than from starvation. The fourfold increase in population there since then might have seen millions die of starvation, as Paul Ehrlich predicted in *The Population Bomb,* had it not been for economic development and global humanitarian assistance.

For all their awesome power, natural disasters cause barely a nick in the rising population graph, or a dent in the world's economy. Compared to deaths from natural disasters, many more people die in road accidents each year: 1.24 million people in 2010 alone. Put another way, the probability of your child dying of natural disasters is down worldwide to a vanishingly small 0.0013 percent.

Disasters are still harrowing, though, when images of every tragedy are transmitted into our homes via television and the internet. As I write this, I am shocked by the death and destruction of the 7.8 magnitude quake in Nepal. I am numbed by the dead bodies being hauled out of the rubble and by grieving relatives wailing uncontrollably. In 2006 I sat aghast watching a mountainous wave pour over the storm barriers of Fukushima and sweep away boats, cars, buildings, and people in a black torrent. I'm riveted and dazed by the impersonal terrifying power of earthquakes, tsunamis, landslides, and floods. I watch, helpless and dismayed, the death and destruction beamed live around the world.

Globally, natural disasters are increasing in frequency yet diminishing in destruction, thanks to a better scientific understanding of the forces of nature, improved early-warning systems, better planning and infrastructure, and national and global preparedness. We owe much of the reduction to the Hyogo Framework for Action, an international treaty guiding disaster relief efforts signed by 168 nations in 2005, aimed at mobilizing international responses to large-scale disasters such as the Fukushima tsunami.

If natural disasters in the past caused the collapse of civilizations and widespread destitution, today, worldwide emergency crews and relief supplies jet in within hours to rescue, feed, and house victims as well as provide financial and technical support for redevelopment in the aftermath.

The origins of modern disaster prevention and responses date to the great Lisbon earthquake and tsunami of 1755, which killed seventy thousand people.[5] King José of Portugal gave the military commander Marquis de Pombal unfettered power to conduct relief operations and rebuild a quake-proof city. Pombal buried the dead in mass graves to prevent the spread of disease, housed victims in temporary huts, distributed relief food, and deployed the army to prevent looting. He handed out questionnaires to compile a damage report used in rebuilding the city, designing quake-resistant buildings and inviting open discussion on the causes of earthquakes. The German philosopher Immanuel Kant chipped in with a naturalistic explanation of the cause of the disaster, paving the way for modern scientific enquiry.

The aftermath of the Lisbon disaster set the stage for the development of modern disaster management, planning, and naturalistic explanations. In the 1940s Gilbert White advanced the science of natural disaster prevention and management by mapping risks, monitoring disasters, and preparing evacuation plans.

In summing up his fine history of disasters, *The End Is Nigh: A History of Natural Disasters,* Henrik Svensen of the University of Oslo shows that the extent of the damage caused by natural disasters, especially among the poor, who are most vulnerable, turns on social cohesion, physical planning, and preparedness. His view is championed by Kenneth Hewitt who, based on comparative studies, points out that death and disruption in the aftermath of a disaster are due more to underdevelopment, inequality, and marginalization than to the magnitude of the event itself.[6]

In Kenya and the Horn of Africa more widely, famine and food shortages in the pastoral lands are, in addition to drought, caused by a breakdown in traditional grazing practices, land loss, and settlement. Data from vegetation plots I've monitored for over forty years in Amboseli show that famine cycles are shortening and tightening. The traditional seasonal grazing practices I documented in the 1960s and 1970s have broken down as land loss, settlement, and a tripling of livestock grazing pressure have reduced the productivity of the grasslands by a third and slowed recovery after droughts.[7]

Going back to my firsthand lessons in Kongwa, the Wagogo's distinction between natural and self-made disasters was a handy one, lost on the colonial agronomists who tried to bulldoze nature into submission. Self-induced disasters are of two types. First, there is the universal tendency to overuse and mismanage our resources because short-term profits outweigh long-term gains. Overuse is the most ancient and universal cause of disaster, dating back to the destruction of the Pleistocene megafauna in Europe, the Americas, Australia, and the oceanic islands. Modern examples include the depletion of two-thirds of the world's fisheries, draining of wetlands, depletion of soils, and overgrazing of rangelands.

Second, there is the Dirty Nest Syndrome I noted earlier, evident, for example, in urban smog and, more covertly, in the pollutants and toxins we dump into the waters and soils. The costs are shocking. The Institute for Health Metrics and Evaluation calculates that eighteen thousand people a day—5.5 million a year—die of pollution-related causes, at an annual cost of $225 billion. Add other welfare costs, including work downtime, and the figure climbs to $5.5 trillion a year. In South Asia, pollution-related losses account for 0.83 percent of the economy.

Such self-inflicted disasters may not be as visible as the Kongwa dust storms, or even recognized as disasters, and that's part of the problem. We pay attention to the immediate and tangible things that affect us because we can see and feel them and tend to ignore the pernicious and invisible impacts mounting slowly and remotely. The divide between natural and human-made disasters is also becoming vanishingly hard to distinguish as our impact penetrates deeper into the Earth's geophysical and biotic systems.

So, for example, the pressurized water pumped deep into the ground to release oil shale in the fracking process is causing earthquakes in Oklahoma to rise from one a year naturally to one a day. In the Sichuan Province of China in July 2013 catastrophic floods and mudslides washed away homes, killing over two hundred people. The cause was put down to pollution emissions reducing sunlight and trapping more heat higher up in the atmosphere, leading to intense rain at night rather than during the day in the nearby mountains. Extreme events long seen as "natural" disasters— heat waves, blizzards, tornadoes, floods, droughts, and dust storms—are all predicted to increase sharply as greenhouse gas emissions warm the planet and cause greater instability in weather patterns.

We have much to learn from traditional societies about disaster awareness and preparedness. Our ancestors learned the hard way to be risk averse, with good reason. Farmers and herders working the land for a living suffer the consequences of bad husbandry practices and learn the rewards of disaster preparation. We moderns typically live far from our sources of food, water, and provisions. We shop for the cheapest product at a supermarket, oblivious to the costs of dollar-a-day labor, the lack of safety regulations, and environmental pollution in producer countries. The disaster in Bhopal, India, in 1984 killed thirty-eight hundred people and led to thirty-eight thousand partial or temporary injuries as a mix of toxic gases used in pesticide production leaked from a Union Carbide factory. In 2013 another disaster hit the headlines when the eight-story Rana Plaza garment sweatshop factory in Bangladesh collapsed, killing over one thousand people and injuring another twenty-five hundred. The immediate cause was attributed to four giant generators on the upper floors causing cracks in the building and the owners flouting building codes to save money. The ultimate cause was a distant price war among department stores in the United States vying for bargain hunters. Americans were inflamed and ashamed by TV scenes of hundreds of dead being hauled out of the rubble and the wailing crowds searching for relatives and friends. A week later, shoppers were back at the local department store bargain hunting.

Nobel Peace Prize winner Muhammad Yunus pointed to the root of the tragedy: "The cracks in the Rana Plaza that caused the collapse of the building [have] only shown us that if we don't face up to the cracks in our own state systems, we as a nation will get lost in the debris of the collapse."[8] His message points to the dark side of globalization: we reap the benefits of fierce price wars and discount the costs to people, the environment, and future generations.

In our global age of such distant unknown and unforeseen consequences, scientists are our risk antennae alerting us to the potential dangers. Often, like scaremongers, theologians, and soothsayers of old, biologists and environmentalists tend toward an apocalyptic view because they project from animal studies and underestimate human ingenuity in redressing the problems we cause. Their blind spots notwithstanding, scientists have dug deeper into the limitations of the environment and looked further into knock-on effects and future implications than anyone else, ever. They look at repercussions far beyond the needs and concerns of any one society, or the horizons of politicians, economists, and planners. Rachel Carson forewarned us of the dangers of chemical, nuclear, and toxic wastes and met a wall of public apathy and the wrath of corporate lobbyists. Today her warnings seem tepid compared to the threat of climate change, acidifying oceans, dead lakes, and endangered species.

Our future is one of a diminishing impact of natural disasters and rising self-induced hazards. The risks we pose to ourselves are far bigger, more entangled, and longer lasting than any in human history. Unless we have the foresight and knowledge, coupled with the necessary monitoring, regulatory, and enforcement capacity, and unless we develop the collective global governance to scale up to a planetary level of protection, our self-made disasters will dim our future. The adage a stitch in time saves nine was never more apposite.

What, then, distinguishes success and failure in avoiding dirtying our own nest as we scale up from living within local ecosystems to the planetary limit?[9]

6

Why Some Succeed Where Others Fail

How do we distinguish success from failure in our management and conservation of the environment? Beyond threadbare survival, how can we ensure that everyone has a chance of achieving a reasonable standard of living and well-being? The answers bear on what we want from the world we've created out of the world we've inherited.

Throughout history we've faced the same recurring problems as Kerenkol—how to make the most of the natural abundance of resources we need to survive and thrive without destroying them. Ecologist Garrett Hardin published a famous article, "The Tragedy of the Commons," in *Science* magazine in 1968, claiming, "Each man is locked into a system that compels him to increase his herd without limit—in a world that is limited. Ruin is the destination toward which all men rush, each pursuing his own best interest in a society that believes in the freedom of the commons. Freedom in a commons brings ruin to all."[1] Hardin's examples of the tragedy of the commons included overfishing the open seas, uncontrolled population growth, poaching, crowding in national parks, urban smog, polluted rivers and lakes, litter, and scrambles for public utilities such as parking lot spaces.

Partha Dasgupta, a University of Cambridge economist renowned for his pioneering work on sustainable development, says of Hardin's "Tragedy of the Commons," "It is difficult to find a passage of comparable length and fame that contains so many errors."[2] Hardin himself acknowledged his mistakes in a later *Science* article: "To judge from the critical literature, the weightiest mistake in my synthesizing paper was the omission of the modifying adjective 'unmanaged.' In correcting this omission, one can generalize the practical conclusion in this way: 'A "managed commons" describes either socialism or the privatism of free enterprise. . . . The devil is in the details.' "

Wildlife is the most ancient persistent victim of our tragedy of the commons. Paul Martin, a tall, gangling paleontologist with whom I partnered in comparing the Pleistocene extinctions in the Americas and Africa, claims the extinction of a giant tortoise in East Africa 1.7 million years ago to be the first evidence of overhunting. Today, these once widespread chelonians survive only on the remote islands of Aldabra in the Indian Ocean and the Galapagos in the Pacific.[3] Martin makes a strong case for a blitzkrieg of human-caused extinctions in the last fifty thousand years.

The argument that the pattern of extinctions mirrors human spread across the continents and around the islands of the world, creating a "deadly syncopation," as paleontologist Ross MacPhee calls it, is compelling.[4] So too is the differential extinction of the largest herbivores, such as the wooly mammoth and giant sloth, and large carnivores, such as the saber-toothed tiger and cave bear in North America. As Darwin noted, the lack of predators on islands like Mauritius made endemic bird species, most famously the dodo, vulnerable to extinction by humans and their retinue of rats, cats, dogs, pigs, goats, cows, and horses. The island of Cousin in the Seychelles is noted for its fairy terns. Elegant birds with snow-white plumage and coal-black eyes, the terns are so accustomed to the lack of predators in their remote island nesting grounds that they will lay their eggs in a fork of a tree and charm you by landing on your head to take a rest.

When I began studying Amboseli in 1967, Hardin's paper was in the works, but the tragedy of the commons was already deeply ingrained in British colonial officers familiar with the destruction of open pastures prior to the eighteenth-century enclosures. Classical scholars trace the tragedy of the commons to the overgrazing of pasture lands in ancient Greece. Aristotle, writing in the fourth century BCE, describes hillsides, denuded by goats, looking like the protruding ribs of a carcass. "For that which is common to the greatest number has the least care bestowed upon," he commented.[5]

Nowhere has the tragedy of the commons been invoked as disruptively as in the British colonial government's push to settle nomadic pastoralists.

Long before conservationists used the same argument to evict the Maasai from Amboseli and create a national park, the colonial administration set up the Ilkisongo Grazing Scheme in the 1940s to control Maasai livestock numbers and movements. The scheme was another Kongwa-like failure. By the time I began surveying Amboseli in 1967, the only signs of the failed scheme were faint lines of thinner bush marking grazing block boundaries. Within a few years I came to a different conclusion than Hardin's. Pastoral systems traditionally managed can be highly productive and resilient, supporting abundant wildlife. I published my findings in science and development journals and, along with social scientist Jeremy Swift and development specialists like Steven Sandford, debunked the view of pastoralism as inherently destructive of land and wildlife.[6] The findings swung the pendulum to the opposite extreme among many anthropologists, echoing the romanticized view of French philosopher Jean-Jacques Rousseau that living in harmony with nature is our primordial condition.

I don't subscribe to Rousseau's view of humans living in harmony with nature any more than I accept that the tragedy of the commons is our natural condition. The tragedy of the pastoral lands over the last half century has more to do with a rising population, breakdown of traditional husbandry practices and governance, the land subdivision and settlement carving up the open range, and the loss of wildlife values.

Few places remain to study large free-ranging wildlife herds, among them the wildebeest in Serengeti, white-eared kob in Sudan, caribou in northern Alaska, and shiru on the Tibetan Plateau. Fewer places remain where pastoralists migrate freely, among them the Dinka and Shilluk in Sudan and the Sami reindeer herders of Europe's northern steppes. The traditional social networks and livestock cultures that once bound pastoralists as a community and anchored them to the land have eroded. I was lucky to begin my studies in Amboseli at a time when pastoralists and wildlife still migrated freely, allowing me to look at how they had survived for millennia without degrading the land or exterminating wildlife.

The earliest formal rules of wildlife ownership were known as *ras nulls* in Roman times, a common law granting the spoils to the hunter who killed or captured a wild beast. The laws did little to protect wildlife from overhunting. The domestication of animals added the specter of unmanaged pastures and degraded pastures.

Hardin considered privatization or strong enforcement by government the only solution to the tragedy of the commons. The enclosure movement in Europe gave evidence of his argument. A dairy farmer fencing her land, managing her pastures, and improving her breeds is the sole beneficiary of her investments and the only victim of her failures. She can make hay in summer and stall-feed her animals in winter, secure a bank loan to improve her dairy production and market access, and buy insurance as a hedge against lean years and disasters. The entire farm operation is within the farmer's control, allowing her to plan and build up a commercial enterprise and pass on the capital gains to her children. East African pastoralists have survived for millennia without either privatizing the land or strong government control, so how did they sustain the productivity of the grasslands and large free-ranging herds of wildlife?

A Maasai herder has access to a huge swath of land and a variety of habitats stretching from arid grasslands to montane forest—provided he builds up a strong network and abides by community customs and regulations. The greater the diversity of landscapes and habitats, the greater the productivity of his herds and resilience to drought and disease. The ecology of scale and social networks in the pastoral lands, like the economy of scale in market economies, creates a win-win for the individual herder and the community.

The intimate links between social networks, the diversity of his herds, the ecology of scale, and his husbandry practices were evident in Kerenkol's success and social standing. His cattle and mixed herds of sheep and goats were more productive and gave him greater resilience to droughts than herders who depended on one species of domestic animal. But a mixed herd calls for many hands and special husbandry skills to meet the needs of each species, separate calves and milking from nonmilking herds, and break up large herds into smaller units better able to feed on sparse pastures. Kerenkol's large social network enabled him to divvy up the herding tasks among family members, kin, associates and, at times, among his several wives in separate homesteads.

When I first started out in Amboseli in 1967, settlements still echoed another fading consideration: defense against raiding neighbors. A settlement brought together an average of twenty households as defense against raiders. In the 1970s when I flew Kerenkol over a group of Ilkisongo warriors gathering to battle the Kaputei Section eighty kilometers to the north, raids were still common, but not for long. Kerenkol leaned out the window and yelled down to the warriors against the battering wind. "Break up! Break up! The police are laying ambush and will shoot if you attack the Kaputei." I suspect it was the first and only time a Maasai laigwenani ever directed his warriors from the air. As the government forces quashed cattle raiding, settlements shrank to two or three households, many to a single family.

A common identity, culture, and purpose: these strong coupled forces explain the extraordinary success of the Maasai in dominating the East African savannas from northern Kenya to central Tanzania. The loosely bound Maasai federation consists of a grouping of clans within fourteen or so large, semi-autonomous sections. Traditional stories trace Maasai roots back to founding families like an oral book of Genesis. The common identity, culture, and purpose binding the federation gives each herder the large social network and extensive territorial flexibility he needs to raise a herd and a family. Within a clan, elders prescribe and monitor grazing regulations and punish rule breakers to ensure the best and most durable use of pasture, set aside grass reserves, and foster unity and peace.

Hardin's point that the freedom of the commons brings tragedy to all overlooks the wild herbivores that also roam freely across the open pastures. Why didn't the great bison herds destroy the American prairies or the 1.5 million herds of wildebeest decimate the Serengeti plains?

The reasons pivot on the same forces keeping the Maasai on the move: the need to shift to better pastures when grazing depletes the value of one patch of grass relative to another. The theory of optimal foraging, as it is known, reminds me of the kid at a party who goes around skimming the icing off all the cupcakes when no one is watching. Animals that ignore the best offerings and stay put as pastures wither lose out to animals on the search for the best forage.

The savanna grasses coevolved with large grazing herbivores and are adapted to withstand heavy grazing for short periods. Grasses cropped and abandoned can replenish leaf growth from root reserves if given time to recover. If grazed persistently, their roots wither and growth halts. The richer, leafier grasses of wet-season pastures are less hardy, dry faster than the coarse grasses of drought refuges, and are abandoned early in the season. Grasses in the Amboseli woodlands and swamps used by wildlife late in the season can draw on more groundwater and nutrient reserves and so tolerate heavy grazing pressure. Abandoned for better wet-season pastures during the rains, the late-season grasses have time to recover before herds return in the dry season.

A landscape broken up by hills, mountains, rivers, and swamps adds habitat variety to the erratic rains and patchy pastures. Each species selects habitats that suit its food demands—elephants the woodier areas, wildebeest the open grasslands, warthog the short grazing lawns. The presence of fifteen herbivore species, each with specialized feeding habits shifting through the tapestry of habitats and periodically disrupted by predators, creates a kaleidoscope of change. Few areas are grazed long enough to degrade the pastures and cause a tragedy of the commons, except in extreme droughts when many herbivores die of starvation. The slackened grazing pressure following a population crash gives the grasslands time to recover in the ensuing rains.

Pastoralists intimately familiar with the land and linked by social networks across their seasonal range benefit from proximity, visibility, and familiarity in making herding decisions. The information the eleenore scouts gather for herders to enable them to collectively decide when and where to move is key to the health of herds and pasture. Wildlife finds its place in the three-thousand-year-old cattle estate of the East African pastoral lands by virtue of its mobility and its value as second cattle in hard times.

Can the lessons of managing the local commons among African pastoralists be applied to the larger global commons? A chance to address the question came when Shirley and I met fashion designer Liz Claiborne and her husband Art Ortenberg on their first safari to Kenya in 1987. Struck by Liz's infatuation with elephants and Art's quick grasp of the part the Maasai played in conserving Amboseli's wildlife and protecting elephants from ivory poachers, I invited the couple back on a conservation safari a year later. At the end of the safari Art and Liz asked if I would help them set up a foundation to find harmony between wildlife and people, as they put it.

Over the next twenty-five years, the Liz Claiborne & Art Ortenberg Foundation assembled a top-rank advisory board and funded conservation programs around the world to save tigers, jaguars, lions, elephants, elephant seals, pronghorn antelope, and dozens of other species. The centerpiece of the foundation's philosophy was the coexistence of people and wildlife. Reviewing reports from the field at the foundation offices on Fifth Avenue in New York in April 1991, we were buoyed by how widely the community-based conservation programs we funded were catching on among international conservation agencies and donors. But was the enthusiasm based on evidence of success, or simply lip service and political correctness to raise funds and appease wildlife conservation critics?

After a couple days spent reviewing reports, we decided to commission case studies around the world and convene a gathering of community members, conservation organizations, donors, government bodies, development agencies, and academics at Airlie House in Virginia in the fall of 1994. John Marinka, a close associate of Kerenkol's and a strong proponent of community engagement in wildlife conservation in Amboseli, joined me on the flight to the meeting. The fifteen case studies, which ranged across community-based forestry, fisheries, wildlife, and protected area and coastal management programs, told a compelling story: conservation practices built around the husbandry of food, water, and natural resources are ancient and deep-seated in subsistence societies—and cooperation and good governance are the hallmarks of their success.

The case studies showed that biodiversity is often conserved incidentally by traditional subsistence lifestyles and cultural practices sustaining forests, wetlands, grasslands, and fisheries. Meg Taylor, Papua New Guinea ambassador to the United States at the time, lamented that highland tribes whose customary practices had conserved some of the richest highland forests on Earth pressed for the right to lease timber concessions to multinational timber companies. The community saw no harm in clear felling the forest, so long as it put money in their pockets. Among the Moluccan Islanders of Indonesia, where customary practices had sustained subsistence fishing communities for generations, the fisheries collapsed when commercial operators moved in and created a free-for-all.

The Airlie House workshop, featured in *Natural Connections: Perspectives in Community-Based Conservation* in 1994, launched a movement giving recognition to community-based conservation.[7] It fell to Elinor Ostrom, a political economist at the University of Indiana, to document and validate the rules for managing the commons on a global scale. As was the case with the Airlie House meeting, the power of Ostrom's work lay in using case studies from around the world, including Swiss grazing pastures, Spanish irrigation systems, Italian communal fisheries, and Japanese public forests. Drawing on dozens of case studies, she defined the Ostrom Eight, as they became known, a set of rules that, taken together, resolved the tragedy of the open unmanaged commons. Ostrom's rules included a strong group identity, collective management and defined boundaries, agreed benefits for a member's contribution, rules governing fair decisions, monitoring of resources and how they are used, punishment proportional to an offense, agreements for broad participation in decisions, and ways of resolving disputes. Even where a strong centralized government prevails, the greatest success in conserving common property resources arises where communities devise their own rules within the larger shared goals of a society.

The beauty and simplicity of Ostrom's classic study, *Governing the Commons: The Evolution of Institutions for Collective Action,* published in 1990, lay in showing how people can and do solve the problem of sharing common resources in practice rather than in theories based on the presumptions of perfect knowledge and the rational actor underlying neoclassical economics.[8] Ostrom won the Nobel Prize for economics for her work, the first woman ever to do so.

My years among pastoralists and the Airlie House findings made reading Ostrom's work a delight after years of battling classical economists who saw privatization as the only solution to resource management. The wonder of Ostrom's work is that all it took to solve the tragedy of the commons riddle was invoking the conservation rules of successful communities based on real-life examples, rather than constructing mathematical theories based on assumptions of how human behavior and culture and societies should function.

Ostrom's rules have been shown to apply across common pool resources ranging from parking lots to public libraries, water use, urban smog, and the planet's ozone hole.[9] In a worldwide review of twenty-five hundred coral reefs, "bright spots," among the most vulnerable of all biomes, showed up with far greater fish abundance than expected based on human population and extraction.[10] The bright spots turned out to be locations where communities still retained strong cultural institutions and customary fishing practices.

What about the persistent failures in conservation? Why doesn't collective action work in the commons more generally if the benefits of cooperation in the community exceed the sum of individual gains?

Our emotions, psychology, and culture each play a large role in what we value and care for. We are better at conserving cows by the billions than restoring the American bison, cats by the millions than the last thirty thousand lions, and exotic plants in botanical gardens rather than in tropical forests—simply because they don't fall within our orbit of concern and daily activities. We domesticate and husband plants and animals that improve our survival, welfare, and well-being. We do well managing resources close at hand and offering quick rewards. We are less successful in conserving things that are hard to husband, degrade slowly, bring deferred benefits, and are costly. Remote and distant threats and benefits seldom show up on our emotional radar or feature high on our list of economic and environmental concerns.

Proximity and tangibility, coupled with strong government policies and regulations, explain why the skies over many industrial cities are getting cleaner, yet the world's atmosphere overall is becoming more polluted, and why the Venetians can raise $8 billion for a barrage to protect the venerable city from sea-level rise, yet we struggle to curb global warming.

Fear and threat also influence our response to other species. We kill snakes and spiders because some are dangerous, locusts because they destroy crops, and lions because they kill livestock. We call in the pest controller to rid our kitchen of cockroaches and rats because we equate them with dirt and disease. And we have no compunctions about eradicating smallpox, malaria, and polio because they are so lethal.

Our sensibilities and understanding of the world also play out in species conservation. Americans are spending millions of dollars restoring the grizzly and wolf after centuries of persecution because they see them in a different light than their grandparents did. Culturally, the world is different from the early nineteenth century, when slavery and child labor were widespread, women were denied the vote, and we were ignorant of our common evolutionary descent with other species. Ecological emancipation, social liberalization, and our modern sensibilities and knowledge have changed who we are. Because the animals no longer pose a threat to us, we are now fascinated by the awesome power of the grizzly and by the intelligence and social bonding among wolves. Where our ancestors saw elephants as meat and danger and drove them to extinction in Europe and North America, we are touched by documentaries of Echo the elephant and her calf in Amboseli and the rangers fighting to protect her from poachers. Culturally, we have evolved into a different species.

So why, despite growing public sensibilities about wildlife, and 196 nations party to the Convention on Biological Diversity of 1992, dedicated to saving the world's biota, are we destroying more species now than in any extinction event since the Cretaceous?

As fascinating as it may be to naturalists, biodiversity simply doesn't grip the public curiosity or attention. There is nothing touchy-feely about biodiversity as there is about an adorable panda, nothing as heart-rending as a baby gazelle struggling to find its first legs and escape a jackal. Biodiversity is beyond our emotional feelers. Conservation biologists have gained ground by conceptually expanding our intellectual sphere to include all life, and the notion is slowly filtering into our language and culture. To narrow the emotional and tangibility gap, biodiversity must have public valence beyond the biologist's deep yet intangible notion of an assemblage of species.

Scientific theories and conservation ideologies, however firmly held, must also be challenged when they fail to explain our behavior and actions. The assumption behind the tragedy of the commons and self-interest trumping the greater public good doesn't jibe with our Good Samaritan instinct to help others in need or to consider the common good.[11] Time and again I found a willingness among leaders like Kerenkol, Parashino, and John Marinka to reach beyond their self-interest for the greater good of their community. Their willingness to give it a go, the buzz we got from working together and seeing the community benefit, the sense of doing good for others—even other creatures—these aren't human traits explained by the selfish gene or kinship theory. Yet they are common human feelings all the same, ones we share and that give life its meaning. Those feelings are strongest in times of disaster. I was struck by how readily San Diegans chipped in to provide food, bedding, and comfort at downtown shelters for tens of thousands of evacuees when devastating fires destroyed hundreds of homes during the drought and fierce Santa Ana winds of 2007.

Where does our willingness to cooperate for the common good come from? How and when does our human cooperative nature work best, and why does it so often fail? How can we expand from the parochial and short-term self-interests of our community and neighborhood to protect the health of our planet and its wealth of life? The answers to these questions will help explain the superdominance of our species and the evolutionary novelty of our conserving other species.[12]

7

Icons of Two Worlds

The sky is cloudless over the craggy Peloncillo Mountains to the north as we saddle up on the Malpai Ranch in Arizona sixteen kilometers from the Mexican border. The six Maasai who have flown in from Kenya and Tanzania are dwarfed by a two-meter-tall American cowboy, Warner Glenn. A renowned mountain lion hunter like the Maasai, Warner is helping them buckle up their chaps for a horse ride on the six-thousand-hectare ranch. They are Joseph Miaron from Amboseli; Yusuf Petenya, Peter Torokei, and Joseph Munge from the Rift Valley on the Kenya-Tanzania border; Dennis Sonkoi from the Loita Highlands near Maasai Mara; and Teteyo Lankio, an Ilkisongo Maasai from Tanzania. The horses shy from the Maasai at first, then settle down when they are soothed and pampered by hands experienced in calming skittish cows.

The meeting of the two cultures is an odd mix—the Maasai among the oldest cattle herders on Earth, the Malpai ranchers the most recent. The Maasai landfall in America was an unnerving moment for Teteyo Lankio, who speaks no English and had never flown before. He is still flustered when he joins us at Glenn's ranch. The plane turned on its side coming into Los Angeles, he tells us. "I thought it was falling into the big water."

In bringing them together from worlds apart, I had in mind lessons the Maasai and Malpai herders could learn from each other. The Maasai were under mounting pressure from the Kenya and Tanzania governments to privatize their land, settle down, and become commercial cattle ranchers. The elders resisted the pressure for years, saying the loss of seasonal migrations would take a heavy toll on their cattle. Lately, though, the swelling numbers of young Maasai entering adulthood with little prospect of building family herds were pressing for private allotments. Nairobi speculators were pushing their case hard, offering the Maasai rock-bottom prices for their land to resell at huge profits to second-homers and developers. Few Maasai realized the impact fencing or the loss of seasonal migrations would have on land, livestock, and culture. The notion of land ownership was so alien that many sold out to the speculators believing they could continue to graze their ancestral lands.

If the Maasai couldn't imagine the impact of year-round grazing on fenced-off lands, I felt sure these ranchers of America's Sonoran Desert could show them firsthand. A land rush in the late nineteenth century following the U.S. Army's defeat of the Apache and Navajo Indians saw the open range of Arizona and New Mexico carved up into sixty-four-hectare allotments—about the size the Maasai stood to get with subdivision of their community lands.

Far from solving the tragedy of the commons, privatizing the open rangelands in the American West had done worse damage. The "tyranny of small decisions," as the impact of a myriad of self-interested goals in creating economic failures and environmental degradation is known, saw homesteaders carve up the open rangelands into ranches far too small to support a family.[1] Impoverished families sold out to land-buying syndicates and venture capitalists who parceled small holdings into big ranches and became wealthy cattle barons. Venture capital and the economy of scale ruled in the cattle industry no less than in any business. Cattle and land became commodities rather than a way of life. Triggered by a booming beef market in the late nineteenth and early twentieth centuries, commercial ranchers overstocked and overgrazed their land. Productive grasslands turned to thickets of mesquite, white thorn, and creosote bush.[2]

By the late twentieth century big corporations had elevated the family ranch to an industrial scale by packing thousands of hulking cattle into small holding pens, feeding them subsidized corn, injecting growth hormones to spur weight gain and heavy doses of antibiotics to prevent disease, and selling off eighteen-month-old steers to supermarkets at cut-rate prices. With feedlot cattle so fat they could barely waddle in the slop of manure, the health of the rangelands and wildlife were not the only victims of industrial beef ranching. The American consumer at the receiving end of the long industrial supply chain was piling on weight from fatty beef, ingesting growth hormones, and fighting infections with antibiotics losing their potency against resistant strains of bacteria due to rampant use in cattle feedlots.

Family ranchers account for less than a quarter of the beef production in America today and cling to their way of life and cowboy culture by doubling up on other jobs. Warner Glenn, for example, earns extra money as a professional guide taking out wealthy clients on sure-footed mules to track down mountain lions in the Peloncillo Mountains. His wife Wendy administers the affairs of the Malpai Borderlands Group. Bill Miller is a commercial pilot checking out electrical power lines for the local utility company. Many families hold down weekday jobs in Douglas, a forty-minute drive away. Like the Maasai's, the Malpai cattle culture is vanishing as the family herd falters, children leave for more rewarding jobs in the city, and ranches are sold off for development.

Wallace Stegner, writer and keen observer of the West, saw the threat to the rangelands and cowboys coming over the horizon and gave this sage advice in 1992: "Angry as one may be at what heedless men have done and still do to the noble habitat, one cannot be pessimistic about the West. This is the native home of hope. When it fully learns that cooperation, not rugged individualism, is the pattern that most characterizes and preserves it, then it will have achieved itself and outlived its origins. Then it has the chance to create a society to match its scenery."[3]

The Malpai cowboys were trying to do just that—overcome their fierce individualism and work together to restore lands degraded by decades of overgrazing by using controlled brush fires and selling conservation easements to stave off land sales to second-homers, dude ranches, and real estate developers. Up in the Peloncillo Mountains the Nature Conservancy had bought the 120,000-hectare Gray Ranch to protect it from subdivision. The Animas Foundation bought the ranch as a grass bank for the Malpai ranchers' cattle. The grass bank serves the Malpai ranchers much as the Amboseli woodlands and swamps do the Maasai: as late-season grazing.

Led by a remarkable band of pragmatic ranchers like Bill Miller, Bill McDonald, and Warner and Wendy Glenn, how had the Malpai Borderlands managed to turn Wallace Stegner's native land of hope and rugged individualism into a novel collaborative movement? To preserve their lands and way of life, the ranchers formed what became known as "the radical center" to counter both the right-wing Sagebrush Revolution, opposed to any government conservation regulations, and left-wing environmentalists behind the "Cattle Free by 93" movement aiming to evict all cattle from federal lands.[4] The mission statement of the Malpai Borderlands Group reads more like an ecological manifesto than a rancher's creed: "to restore and maintain the natural processes that create and protect a healthy, unfragmented landscape to support a diverse, flourishing community of human, animal and plant life in our borderlands region." The man behind the vision, Drum Hadley of the Anheuser-Busch family, preferred cowboy poetry and the open range to the beer business.

The Malpai group linked up with government land and resource agencies, conservation groups, and scientists to preserve native scenery and restore rangelands. Remarkably, in keeping with their new creed, members of the Malpai group began conserving the endangered leopard salamander as well as the threatened horned rattlesnake and prairie dog, species their parents had spent decades eradicating.

I first visited the Malpai ranches in the early 1990s, drawn by the parallels between the Malpai cowboys and the Amboseli Maasai. Both were at risk of losing their land and way of life in battling a deteriorating range. Both straddled some of the finest wildlife lands in the world, and both saw hope in diversifying their income through wildlife conservation and nature tourism.

Over years of mobilizing community support for wildlife, I had found that pastoralists learn new ideas most readily from other herders rather than from government agents and conservationists. I felt sure the radical step of "over the horizon learning," as I dubbed the exchange between Maasai and Malpai ranchers, would work just as well.

In 2002 I had arranged a visit by the Malpai group to meet the Maasai in Kenya with the support of Charles Curtin, an ecologist monitoring the Malpai grasslands.[5] The Maasai were used to hosting wildlife tourists and treated the six American ranchers warmly. Stegner's rugged individuals learned a good deal from the Maasai about cooperation in managing the rangelands and living with wildlife. The return visit in 2004 proved a far bigger challenge. The Maasai will be overwhelmed by America and an alien culture, I was warned. Undeterred, I felt sure the pastoralists and ranchers would bond once they got to talking about cattle and lions, and so it was.

We head deeper into Malpai Ranch interior across mesquite and creosote flats, Dennis Sonkoi and Yusuf Petenya jouncing awkwardly in their saddles, on the lookout for cattle and wildlife. An hour or so later we come upon a few cattle, shy as deer, scattered in the mesquite and yucca scrub. Yusuf turns to me and asks in Swahili: Why are the cows so scared of us? Where are the herders and why have they abandoned their animals for the lions to eat?

The banter about riding a horse and "bush cows" keeps up until we reached the Glenns' ranch house where Warner and his daughter Sarah show us how to rope a calf from the saddle. The Maasai, sure they have a special bond with all cows, try roping the strapping calf on foot and send it bolting over the corral fence. Why do the cowboys keep their cows in the bush and their dogs in the house? With us it is the other way around, they joke.

As we sit on the patio of Warner and Wendy Glenn's house looking out across Malpai ranchlands, the talk turns serious. What can the Maasai learn from Malpai about privatizing Maasailand, fencing, permanent homes, and beef ranching? How can the lessons from East Africa help Malpai ranchers cooperate in setting aside grass banks, burning to restore grasslands, and promoting wildlife tourism to top up ranch incomes? I am struck by how readily the two worlds merge, at how talk of their cattle, family, culture, land, and wildlife transcends continents, ages, race, and ways of life. I am barely needed as a mediator. The pastoralists and ranchers are soon swapping stories about fighting off lion attacks on their cattle, commiserating about droughts, and exchanging poems and songs valorizing their way of life.

In the evening we are guests at a cowboy gathering. Outside the cattle barn the Maasai show the cowboy kids how to throw a spear using a yucca stem. Back inside the ranchers show the Maasai how to barn-dance to the beat of folk guitarist Kip Calhoun. As the evening draws to a close, the Maasai do a sinuous rain dance. Miraculously, it sprinkles the next morning. At a photo shoot at the end of the trip, Wendy Glenn is teary-eyed; this is the most remarkable visit we've ever had, she says.

The exchange was far richer than I had imagined. The most impassioned talk centered on land, cattle, wildlife, and restoring the rangelands. The ranchers told of how they were pushing the U.S. Bureau of Land Management and the U.S. Forest Service to reverse decades of Smokey the Bear fire suppression to let natural fires burn back the scrub as the Indians had done, and how they were enlisting the prairie dogs to chomp down and fertilize grazing favored by cattle and wildlife. The Maasai suggested they go a step further and shift their cattle corrals periodically to create fertile patches of pasture.

Talk of restoring the grasslands to bygone times turned radical when Paul Martin's ideas came up. Drawing on evidence from fossil sites across the southern United States showing stone spear heads embedded in mammoth fossils eleven thousand years ago, Martin put forward a provocative theory that the Clovis hunters, early American Indians who developed sophisticated and lethal projectile weapons, had exterminated the North American megafauna in a blitzkrieg of killings. To return North America to its past glory, he suggested restoring extinct elephants, rhinos, lions, and other species using ecological surrogates from Africa to do so.[6]

Paul Martin had contacted me a year or so earlier about his Pleistocene restoration idea. I had opposed it, pointing out the risks of alien species to America's indigenous plants and the heritage value of Africa's own wildlife. I threw out an idea to the Malpai ranchers I had put to Martin tongue in cheek: if you want to mimic natural processes to restore the land, perhaps you should team up with American zoos and offer elephants kept in barns all winter a sunshine vacation in the Southwest. Zoos could save money by letting elephants roam free on the ranches each winter and provide an ecological service. If nothing else, the thought of elephants breaking up the Malpai thickets got some light-hearted banter going about using ecological surrogacy rather than heavy industrial equipment and chemicals to restore the grasslands.

We had kept the press at bay to avoid the disruption of reporters and photographers eager to cover the glamour of an exchange between the Maasai and the Malpai ranchers. The exception was Verlyn Klinkenborg of the *New York Times,* an astute editor with a soft spot for farming families struggling to hold the line against big business. Verlyn jotted down notes in the background until the final day when, hunkered down in an arroyo over lunch, he threw out a challenge to the Maasai: I've listened; I'm intrigued. But everyone has a story. What should the American public know about your story? What makes your way of life so special? Why is your livestock culture so important to you? How does wildlife fit into your story?

Verlyn captured their story and the cultural similarities with the Malpai ranchers in an erudite editorial observer piece, "A Meeting of Cattlemen from Two Cultures on the Mexican Border," in the *New York Times* on May 9, 2004:

> Beneath the differences—dusty jeans and silver belt buckles versus traditional red-checked blankets and beaded bracelets—lay profound similarities. The Malpai ranchers, most of whom ranch on a modest economic scale, banded together a decade ago to collectively manage their delicate ecosystem. . . . With state and federal agencies, they planned a fire regime that allows natural fires to burn themselves out instead of being suppressed. They have turned this harsh region into a laboratory where scientists can study the natural rhythms of grasslands. The ranchers argue, and science has begun to show, that natural fire and appropriate grazing keep the grasslands healthy and open and that cattle are an environmentally valuable presence. The same issues confront the Masai but in more extreme forms. Over the past few decades their lands have dwindled to the point where some Masai have begun to feel that the only way to hang onto any land at all is to convert it to private ownership. That would be death to a pastoralist culture, whose villages, easily moved in search of fresh grass, become biological hot spots—once they've been abandoned. . . . The knowledge of how to serve these lands, the Malpai and Masai lands, belongs to the people who live on them.

I've been astonished over the years at how readily the savanna pastoralists I've known adapt to change. Despite their seemingly unchanging life, they are constantly testing, modifying, and incorporating new ideas. In the course of the Malpai exchanges the Maasai captured the essence of the situation succinctly: "Your past is our future and our past is your future," meaning the cowboys could learn from the Maasai about cooperation, and the Maasai from the cowboys about tackling land subdivision, degradation, and threats to their way of life.

Warner Glenn, out on a ten-day mountain lion hunt in 1996, struck cat tracks bigger than any he had seen before. Following his baying hounds, he came upon a jaguar, treed and snarling. Warner managed to get off several shots—with his camera. He wrote of the encounter with the first jaguar to return to the United States in forty years in a delightful book, *Eyes of Fire*. "It will take all of our efforts to protect this animal and the wide-open country it needs," he wrote.[7]

Winning space for wildlife is a radical turnaround for ranchers, who a generation ago hunted down wolves as vermin and eradicated all but a handful in the remoter reaches of the Canadian borderlands. The spirit of Malpai soon spread around the Southwest and up through the Interior West to Idaho, Wyoming, and Montana, spawning landowner and conservation organizations like the Blackfoot Challenge and the Montana Wilderness Association, each working to sustain a ranching culture along with the grandeur of the West and its wildlife.[8]

Through the work of their foundation, Liz Claiborne and Art Ortenberg played a key role in galvanizing what became known as the Collaborative Natural Resource Management movement in the Interior West. Following the lead of the Airlie House meeting, the foundation brought together ranchers, conservationists, and government agencies in 2001 at Red Lodge in Yellowstone National Park to promote collaboration in land management in the American West.

The visits between the Maasai and Malpai inspired similar cultural exchanges between pastoralists and ranchers as far apart as India, China, Mongolia, the Middle East, North Africa, East Africa, and South America. The communities share a dedication to conserving a way of life and living landscapes through ecotourism, conservation easements, carbon credits, payment for ecosystem services, and other novel ways of valuing their lands.

Time and again I've been struck by how responsive conservative and deep-rooted cultures are in taking a cue from respected leaders and in exchanging ideas with others who share their values. All it takes are a few respected leaders like Kerenkol in Amboseli and Bill McDonald and Warner and Wendy Glenn in the Malpai to start a chain reaction that can spread through firsthand contact across ranch fences and continents. World-renowned anthropologist Margaret Mead, famous for her study *Coming of Age in Samoa,* captured the role of leadership best: "Never forget that a small group of thoughtful committed citizens can change the world: indeed, this is the only thing that ever has."[9]

What a remarkable turnaround America has seen in a half century: from eradicating to conserving the wolf and jaguar. How sentimental and soft Warner's pioneering forebears would think their grandchildren had become, and how ridiculously altruistic their love of wildlife and dedication to protecting wildlands. Seen in this historical context, how has conservation diversified from its roots in Maasai survival to include a way of life, culture, well-being, and concern for biodiversity at the heart of the new Malpai creed?

The question has nagged me ever since Kerenkol queried my passion for wildlife. It resurfaced all the stronger during Maasai-Malpai exchanges. The change in our values and aspirations in so short a time brings up the deeper mystery of how we arose from our primate ancestry in the savanna to colonize the Earth, and why at the eleventh hour a Warner Glenn should be inspired to write *Eyes of Fire* rather than kill the first jaguar returnee to the United States.[10]

8

An Altruistic Species

By the late 1980s Shirley and I had two young children, Carissa and Guy, who were growing up equally at ease in Kenya and California, where Shirley returned each spring to teach courses on primates at the University of California, San Diego—continents and a world apart from where I grew up in Tanzania in the 1950s. Yet these two ends of the human diaspora, originating in the Old World and reaching a pinnacle of technological mastery in the New, were rapidly converging on a common global economy.

California had changed since my first visit in the early 1970s. Shirley had been a part of the civil rights and environmental movements at Berkeley at the time, marching for social justice and the environment. Similar marches across America helped the white visage of college campuses diversify to include African Americans, Latinos, and Asians. Thanks to the Clean Air Act of 1970, the skies over Los Angeles and San Diego were cleaner and the bald eagle, grizzly bear, and wolf were making a comeback after the passage of the Endangered Species Act in 1973.

Notwithstanding the cleaner skies of Southern California and the rebound in wildlife, the United States today faces a new and far more profound challenge. The oil crises that sparked tumult at the gas pumps in the 1970s and prompted legislation to raise vehicle efficiency and cut freeway speed limits to eighty kilometers per hour gave way in another two decades to a Me Generation more interested in driving jacked-up, gas-guzzling ORVs than subcompacts. Big corporations were shifting production to offshore sweatshops in Asia, where South Korea, Thailand, Malaysia, India, and China were chipping away at America's economic dominance. Fears of the 1960s population explosion were waning, but with the developing world on a fast track to American prosperity, the coal and oil power stations fueling the world's economies were venting ten times the pollutants emitted by the United States at a similar phase of development, raising early concerns over CO_2 emissions and global warming. In Mexico, a growing population and sluggish economy were driving waves of job seekers across the border into America.

Watching the impact of the newly emerging economies surge ahead, I worried that the political commitment to the principles of sustainable development expressed at the World Commission on Environment and Development in 1987 was no more than lip service. How could biodiversity survive if the emerging nations followed the same "Develop now, clean up later" path as the industrial West?

A chance to explore the question came when Bill Conway and George Schaller asked if I would take over directorship of the Wildlife Conservation Society's international program. George was eager to hand off the responsibility and focus on his panda work in China after his pioneering work on gorillas, lions, tigers, jaguars, and snow leopards. The society's conservation scientists at the time were a stellar bunch. Aside from George and Bill, there was Roger Payne, working on humpback whales, Tom Struhsaker, working on primates in Uganda, Charlie Munn, studying parrots in Peru, Amy Vedder and Bill Weber, researching gorillas in the Virunga Mountains in Rwanda, and Archie ("Chuck") Carr, noted for his turtle work. The Wildlife Conservation Society (WCS) program was also attracting talented young field biologists, among them Alan Rabinowitz, working on jaguars in Belize, Lee White and Mike Fay, concentrating on Central African rain forests, Graham Harris, working on wildlife in Patagonia, and Ullas Karanth, focusing on tigers in southwest India.

Seeing an opportunity to draw on their talents, I convened several meetings to launch a WCS international program to protect forests, grasslands, mountains, and coral reefs around the world. I envisaged the program creating regional centers in the developing countries to train a cadre of nationals, and I took the first step, setting up the African Conservation Centre (ACC) in Nairobi. In the coming years ACC would train dozens of African nationals from as far afield as Nigeria and Malawi and launch an independent African conservation hub.

The remarkable dedication to conservation among field biologists at WCS drew me back to Kerenkol's bafflement over my own passion for wildlife. Why should such a talented group of scientists devote their lives to saving rain forests, whales, and jaguars in remote places as if it were the natural thing to do, as natural as Kerenkol raising his cattle and a big family?

Nothing in our evolutionary past hints at our radical change from exterminating and displacing other species throughout our rise to global dominance to saving them from extinction today. Charles Lyell, scion of nineteenth-century science, put our rapacious past starkly in his 1833 *Principles of Geology*: "In thus obtaining possession of the Earth by conquest, and by defending our acquisitions by force, we exercise no exclusive prerogative. Every species which has spread itself from a small point over a wide area, must, in like manner, have marked its progress by the diminution, or the entire extirpation, of some other, and must maintain its ground by a successful struggle against the encroachment of other plants and animals. . . . We may regard the involuntary agency of man as strictly analogous to that of the inferior animals. Like them we unconsciously contribute to extend or limit the geographical range and number of certain species, in obedience to general rules in the economy of nature, which are for the most part beyond our control."[1]

Lyell's treatise, showing that the Earth was formed over eons by the slow persistent action of upheavals and weathering, profoundly influenced his protégé Charles Darwin as he delved into the evolution of life during his voyage on the HMS *Beagle* a decade later.[2] Oddly, Darwin, though horrified by the slave trade, made no mention of the extermination of wild animals by commercial whalers and meat and hide traders, nor of the European farmers and herders erasing the forests and native habitats of South America, Australia, and Southern Africa. The extinction of species wasn't even recognized scientifically until the French naturalist Georges Cuvier described the fossil remains of giant mammoths in 1796.

Building on Cuvier's discovery, Lyell and the budding science of geology unearthed compelling evidence of past ages dominated by large creatures long vanished from the Earth. Darwin, familiar with Cuvier's work, drew on the evidence of extinctions in deducing his theory of evolution.[3]

Like Cuvier, Darwin attributed the demise of extinct species to climate swings. Thomas Huxley, an eminent biologist and ardent defender of Darwin's theory of evolution by natural selection, delivered an address at the 1883 International Fisheries Exhibit in London dismissing any role of humans in extinctions: "Any tendency to over-fishing will meet with its natural check in the diminution of the supply, . . . this check will always come into operation long before anything like permanent exhaustion has occurred."[4]

Alfred Russel Wallace, with Darwin the first to articulate the theory of evolution by natural selection, said of the extinctions: "We live in a zoologically impoverished world, from which all the hugest, and fiercest, and strangest forms have recently disappeared." He thought us better for their extinction: "It is a much better world for us now they have gone." He attributed extinctions to past climatic changes. Only late in life did Wallace change his mind. In his last book, *The World of Life: A Manifestation of Creative Power, Directive Mind and Ultimate Purpose,* published in 1911, he wrote: "The extinction of so many large mammalia is due to man's agency. The whole thing is really very obvious."[5]

What caused the sudden shift in our views, from Darwin's blindness to the role humans played in the extermination of species to Wallace's revelation and Conway's enlarged sensibilities to conserve all species everywhere? What explains the change from acceptance of the biblical injunction to subdue nature in the nineteenth century to saving biodiversity a century later? How did we escape Lyell's general rule of nature that he thought was intrinsic to all creatures and beyond our control?

E. O. Wilson, an eloquent voice of the conservation biology movement, says that human destructiveness is something new under the sun.[6] Lyell was closer to the mark in saying we are no different than other species. What really makes us aberrant, and only so in modern times, is reining in our reproduction, recognizing the limits to growth, reaching beyond kin to consider other peoples and, above all, saving other species. This evolutionary anomaly, this break with a gene calculus that extends our circle of empathy to the rights of other tribes and other species, sets us apart from all others. Saving other species moves us from viewing humans as the divine object of creation to an evolutionary and ecological view in which we are one of 10 million species enmeshed in a web of life. Conserving other species for their intrinsic value goes beyond our evolutionary self-centeredness and gratification, beyond race and religion, to the highest pinnacle of altruism—saving all species for all time.

Is altruism—helping others with no expectation of reward and even at a cost to ourselves—anything more than extended self-interest? Biologist David Sloan Wilson in *Does Altruism Exist?* probes this question in exploring the evolution of consideration extended beyond our circle of relatives.[7] Altruism was one of the big mysteries Darwin grappled with in the *Descent of Man* in trying to explain the sacrifices social insects such as worker bees make in forgoing their own reproduction to feed the queen's young. Surely, he worried, such evident self-sacrifice undermined his theory of natural selection. Noting that people do risk their lives for others, he suspected the roots of altruism lay in the human capacity for empathy.

J.B.S. Haldane anticipated the modern genetics view of kin selection in addressing Darwin's dilemma. A British population geneticist and polymath with an eclectic scientific appetite and compassionate political views, Haldane moved to India in later life and became an Indian national. "I would sooner rescue two brothers from drowning than two strangers," he said in an offhand remark. His genetic take on altruism was formalized as "kinship altruism" and "inclusive fitness" by Oxford biologist W. D. Hamilton and popularized by E. O. Wilson in *Sociobiology*.[8] Inclusive fitness theory holds that a species' success depends not only on the number and survival of its own offspring, but also on the number of genes its relatives contribute to the gene pool.

According to Hamilton, social insects were the exception that proved the rule. The worker bees are all progeny of the queen dedicated to and specialized in nurturing her offspring and enhancing her inclusive reproductive success. Richard Dawkins took the fitness theory a step further in his 1976 book *The Selfish Gene*, arguing that all selection is a battle between genes to perpetuate themselves.[9] Cooperation, in his view, amounted to no more than genes using other genes to advance their frequency in the population.

Social scientists, railing against the genetic reductionism of sociobiology and insisting there is far more to altruism than kin selection, countered the selfish gene view by citing cooperation among nonrelatives and the role of culture in influencing behavior in social evolution. Cooperation is the very essence of cultures, making governance of societies and civilizations possible. Genetic evolution occurs far too slowly to explain cultural evolution, they insisted. Many biologists shared their skepticism. Ernst Fehr and Simon Gächter stated, "Human cooperation is an evolutionary puzzle. Unlike other creatures, people frequently cooperate with genetically unrelated strangers, often in large groups, with people they will never meet again, and when reproductive gains are small or absent. These patterns of cooperation cannot be explained with the evolutionary theory of kin selection alone."[10]

Even Dawkins admitted to being baffled by humans. In *A Devil's Chaplain* in 2003 he stated: "What I am saying, along with many other people, among them T. H. Huxley, is that in our political and social life we are entitled to throw out Darwinism, to say we don't want to live in a Darwinian World."[11] In an open letter to Prince Charles Dawkins makes a far more relevant remark about the unique capacity of the human mind and culture: "Long-term planning . . . is something utterly new on the planet, even alien. It exists only in human brains. The future is a new invention in evolution."[12]

If Dawkins accepts that human altruism is too good for our selfish genes, how can we explain general altruism and its recent expansion to include other creatures genetically unrelated to us?

When theory fails to explain human behavior, it speaks to simplifications scientists make rather than the failure of the scientific method. Modeling the complexity and nuances of human behavior calls for more data than we can gather and computers handle. To get around this dual problem, mathematical modelers pare down the number of variables and simplify their assumptions. Scientists who've spent years on theories of human behavior are loath to admit failure and come up with contingencies to explain exceptions—or they throw up their hands like Dawkins has and declare us a Darwinian exception.

Where, then, does the impulse to save other species come from? Is it genetic, and so instinctual, misplaced empathy, or it is learned and therefore cultural behavior? E. O. Wilson explains our concern for other species by assuming we have an innate tendency to focus on life and lifelike processes. "Biophilia," as he calls it, compels us "to look to the very roots of motivation and understand why, in what circumstances and on which occasions, we cherish and protect life." He goes on to say that biophilia reaches beyond material and physical sustenance to human craving for aesthetic, intellectual, cognitive, and spiritual meaning and satisfaction. "To the extent that each person can feel like a naturalist, the old excitement of the untrammeled world will be regained."[13]

As a naturalist I agree with Wilson's excitement about nature, but our views are not endemic to the human species. There is little evidence to support an intrinsic love of nature. How, after all, does biophilia square with Wilson's claim that human destructiveness is something new under the sun? We can't have it both ways. A concern for biodiversity is an outcome of human knowledge and sensibility still largely confined to more educated and wealthier societies, and largely alien to those struggling to survive and conquer nature.

In a scientific review of Wilson's biophilia hypothesis, Stephen Kellert, though an advocate of the biophilia hypothesis, concludes that most Americans and Japanese express concern for only a limited number of species and natural objects. Jared Diamond in the same review points out that concern for animals revolves around ownership of useful species.[14] There is little evidence of species being viewed as community property and abundant evidence showing that our history is replete with overkills, as Paul Martin claims.

Drawing on my own experience, how do I explain my transition from a love of hunting to an abhorrence of killing animals, from wanting to conserve wildlife in my neck of the woods to working to conserve biodiversity globally? I trace my transformation to the same forces driving the emerging notion of global stewardship—the expanding awareness, knowledge, and sensibilities arising from my life's experience and new cultural values.

Although I would like to think my passion for animals is a latent propensity left over from our savanna origins and common to all people, I see saving other species as something truly novel in evolution—a unique, hard-won trait of modernity still exceptional and tenuous around the world. Saving other species emerges in our emancipation from nature—in our expanding circle of sensibilities to include others beyond family, tribe, and ethnicity. To understand how biodiversity conservation can be nurtured—how we can buck the evolutionary impetus to reproduce and displace other species—we need to know where conservation itself came from, and how it evolved out of our own species-centered origins in the savannas to saving species around the world. Is conservation unique to our own species, and if not, does it originate in other species? If unique, what explains the origins and evolution of conservation from Maasai survival strategies in the 1970s droughts to my passion for other species? Only by answering this enigma can we say whether we are doomed to hew to Lyell's economy of nature's relentless drive to reproduce and outcompete other species or if we have developed a novel cultural urge to conserve other species. If a novel culture, can we speed up the urge, or are we too constrained by our ancestral behavior and the materialism of the modern age to avoid other species' destruction and perhaps our own too?

My interest in these deeper questions about conservation began with expeditions I took to Jordan, Syria, and Lebanon in 1965 and, in the following year, to the Sahara Desert along the Moroccan-Algeria border. Here, in the hottest places on Earth, I was puzzled at the remarkable ability of desert animals to survive in the searing heat and bone-dry deserts and set out to explore why.

Over the course of two summers I logged the daily activity of agama lizards using a multi-channeled temperature recorder. Baking in fifty-five-degree-centigrade heat, I monitored the body temperature of agamas and the ground and air temperatures from their first emergence from rock shelters in the early morning to the torrid midday hours. The lizards appeared as soon as the sun warmed the ground, flattened themselves against the hottest rocks, and absorbed the heat through their darkened skin. Once warm, the lizards darted across the hot sand grabbing insects, then scurried back to the coolest and lightest colored rocks, switched to a pallid color in chameleonic fashion, turned head-on to the sun to minimize its effect on their bodies' surface area, stood high on their elongated toes to minimize heat absorption, and later retreated to the cool recesses of rocky outcrops in the blasting midday heat. Using the sun to raise their body temperature, avoiding the withering heat, imbibing the water of their insect prey, recycling water physiologically, and venting dry pellets, the lizards conserved enough energy and water to survive the extreme desert heat and aridity. The Bedouins who called by wondered why I was studying lizards to find out how to survive the deserts when they could tell me firsthand. They had a point.

Most animals have evolved elaborate ways to conserve energy and water, whether polar bears storing body fat and hibernating through winter in the Arctic or desert tortoises storing water in their urinary bladder. As a savanna species, we also evolved anatomical and physiological adaptations to conserve energy and water. We build up fat in good times and turn down our metabolism when food and water run short. The ghost of our evolutionary past haunts us when famine is replaced by the excess calories of modern living, as I indicated earlier. When we diet to shed excess weight, our metabolism drops like that of Maasai zebu cattle in droughts in order to conserve energy. Stop dieting and we put on weight faster than ever. We are adapted to a food surplus in good times, not all the time.

Extrasomatic conservation of food and water is far rarer than bodily storage in animals other than us, but common all the same. Squirrels bury nuts and beavers store harvested plants in their nests as a hedge against winter shortages. Leopards kill excess gazelle and hang them in trees like laundry to eat when the herds move on. Some animals are particularly smart about surviving food deprivation. I couldn't fathom how hyenas thrived after eating their way through the thousands of carcasses left after a severe Amboseli drought in 2009. The answer came when I discovered hyenas soaking mummified carcasses in ponds to soften them enough to chew and swallow. The most remarkable example of storing food is found in the jay family. The scrub jay of California stores up to twenty thousand acorns in preparation for a winter deficit and remembers where most of them are with a prodigious spatial memory unrivaled among humans.

These examples show that animals are genetically and behaviorally primed to build food stores and conserve water against hard times. In this sense they are practicing conservation for survival, much like our hunter-gather forebears. Yet conservation among other species falls far short of the practices of early farmers and herders who raised and sustained the productivity of their natural foods through domestication and husbandry. The social insects offer something of an exception, though. Termites raise and sustain the productivity of their food supply by domesticating and farming fungi and aphids. Ants on the island of Fiji have in the last 3 million years evolved a far more elaborate husbandry system, in which they inject the seeds of six species of plants into the bark of trees to germinate them, then fertilize them with their feces. The plants are dependent on the ants for propagation and the ants on the plants for food and shelter.[15]

Despite their limited cognitive capacity, social insects alone among species other than us have mastered domestication and farming. E. O. Wilson, in *The Social Conquest of Earth,* argues that social insects and humans are the most successful of animals because of their ability to cooperate as a super-organism and put the interests of the group above the individual.[16] Yet for all their cooperative skills, the social insects are gene-bound and lack the ability of humans to reach beyond kin to cooperate at a global scale and reshape ecology and evolution.

We can ask the same question of termites that I asked of elephants. If they ruled the world, would they be concerned about saving humans? No, they would fail the test of altruism beyond their own species just as dismally. After all, the notion of biodiversity is not imbedded in our genes; it is a cultural outcome of our scientific discoveries of how we came to be as a species, and why there are so many besides ourselves.

When Venetians conserve their past and Tanzanians set aside the Serengeti for future generations, when nations aspire to protect the diversity of life for its own sake, we stand alone among species. We have broken the evolutionary rulebook and risen beyond the inborn urge to survive and favor our own kin, tribe, race, and species—we have taken altruism to a new plane. Although we share with other species the conservation of energy and resources, the scale, scope, and diversity of things we conserve and our reasons for doing so have no evolutionary precedent. Our notion of conservation calls for extraordinary cooperation: forging a common set of goals and planning far into the future. Our rise to global conquest took a path that not only gave us an edge in competition with other species but, in vanquishing them, prompted a belated propensity to save them from our own dominance. What made us different, and where did our concern for the future of life and our planet come from?

9

Breaking Biological Barriers

I had the good fortune to meet and marry Shirley Strum. A Berkeley graduate tutored by Sherry Washburn, whose teachings broke down the barriers between anatomy, physiology, and genetics to usher in the new physical anthropology, Shirley was studying savanna baboons when I met her in 1973. She had begun her research into baboons as models of early human evolution in the savannas a year earlier on Kekopey Ranch, located among the buckled fault-scarps of the Rift Valley in central Kenya. The troops she followed on the ranch ranged across the Kariundusi Stone Age site where early hominids had fashioned Acheulean stone tools a million years before.

Shirley's study coincided with the arrival of a new baboon male, Ray, who was trying to inveigle his way into the troop. According to Sherry Washburn and Harvard anthropologist Irv DeVore, baboon society was built around a male dominance hierarchy. The strongest male stood at the apex, with subordinates in a descending pecking order. Shirley fully expected the newcomer to fight the resident males over females.

Ray did no such thing. Instead he approached females with solicitous gestures and gradually befriended them, and through them, their infants. Ray was using stealth, solicitation, and appeasement to lower tensions and ease his way into the troop. Through her study Shirley discovered that male baboons use social smarts as well as aggression to boost their reproductive success in winning over females.[1] A female threatened by an aggressive male tends to avoid him in favor of a friend who protects her infants. The smart male hugs compliant infant friends to ward off aggressive males, knowing the rest of the troop will mob the attacker if the infant screams. A smart male, in other words, makes headway by building alliances and a social network. He also benefits from forming friendships, which lowers stress and reduces the risk of severe injury in fights over females.

Ray's new troop was called the Pumphouse Gang, named after both the California surfer group made famous by Tom Wolfe and the location of the ranch wellhead in the middle of the troop's home range. Through Shirley's work, Pumphouse became a star of *National Geographic* magazine and films, portraying the social sophistication of baboons and countering the imagery of the aggressive, warfaring primate popularized in Robert Ardrey's best-selling books *African Genesis* (1977) and *The Territorial Imperative* (1966).

In her own popular book, *Almost Human,* published in 1987, Shirley showed how remarkably sophisticated baboons are. Her decades of work, first at Gilgil in the 1970s and then on the Laikipia Plateau, where she moved her study troops in 1984 to save them from the spread of farms, painted a new view of primates. Rather than the fixed-action automata depicted by animal behaviorists at the time, Shirley showed baboons to be socially smart and, like Ray, adept at building friendships and alliances. Social smarts stretched the here-and-now calculus of aggression to long-term investments in relationships that pay off months and years ahead. Her studies challenged the selfish gene view of primates and echoed Charles Darwin's dictum in *The Descent of Man*, written more than a century earlier: "He who understands baboons would do more towards understanding metaphysics than Locke."[2]

When our own budding friendship led to romance and romance to children, Shirley introduced Carissa and Guy to her study troops. She was immediately impressed by the remaining gulf between baboons and humans, though she had done so much to narrow the gap. Carissa, face-to-face with baboons, wanted to befriend them. The baboons, to the contrary, had no interest beyond curiosity in the odd homunculus. Shirley, an inveterate primatologist, recorded how Carissa and Guy at an early age reached out to other people and animals, and how readily they made friends with strangers, offering them food and toys and sharing things. They learned fast by closely following Shirley's every action and gaze, imitating the finest details. When Carissa and Guy began talking, Shirley recorded their babbles, budding sounds, and baby language as minutely as she studied baboons, but eventually gave up: it would have taken a full-time child psychologist to document the welter of innovations and permutations. Besides, our kids were at an age to take umbrage at being compared to baboons.

Carissa and Guy switched Shirley's attention from how baboons were almost human to why they were not human. What, she asked, stands in the way of baboons becoming as smart and dominant as humans? The higher primates have evolved a rich repertoire of cooperative behavior, so why have they not developed our breadth of cooperation and evolved farther along the road toward us?

The evolution of social skills is the engine of intelligence in primates, according to psychologist Nick Humphrey.[3] The discovery of complex social skills led Shirley's coworkers Andy Whiten and Dick Byrne to dub primate savviness "Machiavellian intelligence."[4] The term casts primates as exploitative and deceptive, whereas Shirley finds baboons more complicated socially. Still, at least scientists were finally acknowledging primates to be intelligent, sentient animals after generations of casting them as reflex-bound creatures. So why haven't other smart primates developed the cooperative hunting and gathering at the root of our rise to ecological dominance?

Shirley's interest in the development and limitations of primate cooperation was first sparked by observing the male baboons who began joint hunting forays on the growing population of Thomson's gazelles on Kekopey Ranch. Cooperative hunters did better than loners. For all their success, though, their cooperation was transient, diluted by the continual emigration and immigration of males between troops.

Closer cooperation does develop in some chimp hunting bands and may extend to nonrelated troop members, but it falls far short of the level of cooperation among large bands of unrelated human hunters. Humans' use of language to plan and coordinate hunts as well as their ability to accumulate and transmit knowhow through successive generations, invent weapons for specialized purposes, and track down and pursue quarry for hours and days at a time without distraction put them in a league of their own.

As Shirley sees it, the behavior of baboons and other primates is like a jumble box. Feeding, sex, nurturing, socializing, and sizing one another up all intersect when animals meet. Although humans have a similar jumble of impulses, we typically separate feeding, working, teaching, play, and socializing. We parcel our time, divide up tasks, and focus on one thing at a time. Dividing tasks and honing our skills laid the foundation of modern industrial societies, as Adam Smith explains in his famous *Wealth of Nations.*[5]

Baboons took the same pathway from forest to savannas as our *Australopithecus* ancestors. Like humans, baboons have thrived as ecological generalists. They are far more numerous and widespread than the smarter great apes—gorillas, chimps, and orangs. In Africa baboons are more abundant than any large mammal except us, thriving in humanized landscapes from ranches to farmlands and in residential areas across most of the African continent. Baboons often get the better of us even in the heart of our own domain. Shirley was invited to Cape Town to help the city figure out how to deal with hairy urban bandits accosting homeowners and breaking into their kitchens and fridges.

Successful as baboons are, they number on the order of 5 to 10 million in Africa, whereas human numbers are close to 1 billion. We achieve densities tens of thousands to the square kilometer in cities; for baboons it's rarely more than thirty to the square kilometer even in their richest habitats. We occupy all continents and every biome on Earth, baboons only Africa and the Arabian Peninsula. We make our own shelters, no longer depending on the cliffs and trees baboons climb for safety at night. We manufacture our own food and habitat. Baboons rely wholly on natural products, except for raiding our crops and fridges.

What, then, explains our extraordinary ecological success? The conventional argument is bipedalism, a big brain, and tool use. All are important and work in tandem. Bipedalism freed our hands to become more dexterous in making tools, and our arms and shoulder articulation allowed us to wield weapons that our big brains figured out how to fashion and handle adeptly. These distinctive traits expanded the variety and size of animals and plants we could harvest and process.

Hunter-gatherer societies range across the open wildlife-rich savannas by cooperating in building sturdy shelters to ward off predators, keep their children safe at camp, and insulate themselves against rain, heat, and cold. They create a home base, allowing them to camp anywhere and move when conditions change. The division of labor allows hunters to forage over a far greater range than other primates and carnivores, and to breach the confines of ecosystems. In doing so, humans outdistance and outmaneuver other species and avoid close competition for resources.

Despite the advantages they give us, bipedalism, big brains, and tools are not enough to explain our ecological dominance. A big brain merely outsmarts smaller ones to the same selfish ends. Chimps are larger brained than baboons, more bipedal, and use tools, but in evolutionary terms they are far less successful. Chimps can outsmart one another and wage war on neighboring troops, but they are stymied by kinship ties among males from forming large unrelated social networks—and never develop the division of labor seen among Hadzabe and !San hunter-gatherers. All nonhuman primates fall far short of the cooperation it takes to harvest and store food collectively in anticipation of hard times. They lack the capacity of humans to ensure that group members reap the rewards of their food gathering, storage, and sharing, with little risk of being thwarted by outside raiders and inside cheats.

The same goes for tools. Macaques and capuchins use conveniently shaped stones as tools. Chimps go a step further in fashioning sticks to nab prey such as termites and bush babies. Despite 7 million years of evolution since the hominid lineage split from the apes, the apes have advanced little in making tools, expanding to new habitats, or tailoring their environments to suit themselves. Termites and beavers do better than apes in engineering habitations for themselves—in niche construction, as ecologists call it.

So long as our early ancestors behaved like other primates, they posed no serious threat to savanna herbivores and carnivores. That all changed when our lineage began hunting.

The oldest known stone tools date from 2.6 million years ago, when brain size had expanded from a chimp-sized 450 cm² to 600 cm² among the fully bipedal australopithecines. Around the same time the first hominids emerged on the scene with a precision grip that allowed them to craft cutting, scraping, hammering, and chopping tools and to up their prey size from small to medium-sized animals. Propelled by these advances,

Homo erectus ventured out of Africa into Eurasia around 2.2 million years ago. In its northern reach, *erectus* encountered colder climates, requiring warmer clothing and more substantial shelters. Though far more successful than the australopithecines in their spread and abundance, these early intercontinental hominids and later emerging forms, including Denisovans in northern Europe and the Neanderthals who spread across Europe from Africa perhaps seven hundred thousand years ago, had little impact on the large mammal community and didn't climb far up the ecological dominance ladder.[6]

Around five hundred thousand years ago, though, something changed, marking a rapid transition within 150 millennia from the large tear- and pear-shaped Acheulean butchering, skinning, cutting, and digging tools—manufactured by *Homo erectus* for over a million years with little change—to a far more refined tool kit. The story is told in the archeological digs at Olorgesailie eighty kilometers southeast of Nairobi in the Rift Valley. Here, Rick Potts, Alison Brooks, and John Yellen of the Smithsonian Museum, with their colleagues from the National Museums of Kenya, began excavations over thirty years ago. In 2014, out on the digs in the sweltering heat of the Rift Valley, Alison explained to me the marked shift from clunky, lightly worked Acheulean tools in the Lower Stone Age half a million years ago to the sophisticated and varied tool kit characterizing the Middle Stone Age. The new kit was made up of obsidian and chert blades, hammer stones, micro flakes, and flensing tools for scraping meat off hides. By then early hominins, perhaps *Homo erectus* or Neanderthals or both, were killing large mammals with projectile spears. Digs in three-hundred-thousand-year-old clay sediments in Schöningen in Germany around the same period unearthed well-fashioned spears made of spruce and pine, along with dozens of fossilized horses preserved in an anoxic bog.[7]

At Olorgesailie, fossils of an elephant, *Elephas ricki,* and hippo, *Hippopotamus gorgops,* both far larger than their modern counterparts, do show signs of butchery as early as 800,000 years ago but there is no evidence of significant extinctions until the transition in tool kit at the end of the Lower Stone Age. Then, between 500,000 and 350,000 years ago, there is a convulsive turnover of large mammal species, including the larger elephants, hippos, baboons, bovids, sheep, and pigs, marking the emergence of our modern African fauna. Whether the turnover in species was the work of better hunters using superior weapons, climate change, or both is still under investigation. Rick Potts has drilled a core through lake sediments and is currently checking the results for indications of a climate shift. Given the wide-ranging movements of elephant, buffalo, zebra, warthog, lion, leopard, and other large mammals in Africa today, and the wide variety of habitats they traverse in the day's walk from the arid rift valley floor to the forest escarpment, I can't imagine a climatic shift large enough to cause such a faunal tumult. At a meeting convened by Rick Potts at the Smithsonian in October 2017, I joined the review team looking into the likely causes of the Olorgesailie extinctions. Although the drill cores showed the climate becoming drier and more varied during the faunal turnover, there was no smoking gun. To me the extinctions bear the hallmark of the human extirpations that have followed our expansion into every continent and onto every island from the first emergence of modern humans out of Africa. Since then the selective loss of large mammals has coincided with our spread and acceleration in colonizing the world.[8]

We do know that the obsidian and chert cores used to fashion the sophisticated tool kit were imported from sites several days' walk away. The delicately knapped tools and sharp cutting edges called for forethought, planning, and specialization. Alison Brooks believes that such skills, coupled with evidence of long-distance travel and perhaps trade in stone cores, required language. Daniel Everett, drawing on archeology, anthropology, neuroscience, and linguistics, believes that language had already emerged in *Homo erectus* more than 2 million years ago, far earlier than the 100,000 years ago linguists previously ascribed to modern humans.[9] Our peripatetic tool-making ancestors may already have developed symbolic markers far earlier too. Red ochre unearthed at the Olorgesailie beds of 250,000 years ago points to the use of ornamentation by then, perhaps to distinguish group affiliations in the same way Maasai use ochre today. Geometric shapes on freshwater mussel shells found in Java and attributed to *Homo erectus*, though still in dispute, may push back our ancestors' use of symbolism to half a million years.

The Olorgesailie story places the emergence of language, culture, and symbolism far earlier than the 200,000-year benchmark paleontologists use to date modern humans, based on fossils found at Omo-Kibish in Ethiopia. Recent studies of fossil hominids found at Jebel Irhoud in Morocco show essentially modern humans at 315,000 years ago, associated with flint blades and wooden shafts likely used for projectile weapons. Similar flint blades of the same age found across Africa point to the wide distribution of our forebears at the time, and to a broad evolutionary front rather than emergence from a single location.

There is a good deal of debate over the cause of our rapid increase in brain size over the last million years: social or environmental? Robin Dunbar of Oxford University has shown a close link between brain size, frontal lobe development, the density of white matter, and primate group size.[10] He suggests that larger groups trigger more social interactions and select for a bigger, more complex brain. Untangling cause and effect from correlations is difficult, though, especially when group size itself is closely tied to food and predator abundance. Shirley and I, based on her study of baboons and mine of large mammal ecology, concluded that both social and ecological factors play a role.[11]

Michael Tomasello of the Max Planck Institute for Evolutionary Anthropology in Leipzig speculates that ecology was key to human evolution and our ability to cooperate: "Early humans were forced by ecological circumstances into more cooperative lifeways, and so their thinking became more directed towards coordinating with others to achieve joint goals or even collective goals. And this changed everything."[12]

Chimp researcher Frans de Waal agrees: "Human psychology evolved to permit ever larger and more complex hunts, going well beyond anything in the animal kingdom. While the actual hunting of large prey may have driven this evolution, our ancestors engaged in other cooperative ventures, such as communal care for the welfare of young, warfare, the building of bridges, and protection against predators. They benefited from cooperation in myriad ways."[13]

Psychologists Martin Seligman and colleagues in *Homo Prospectus* go further still: "The deadliest predator on the planet is not the strongest or swiftest, but the one with the longest time horizon of anticipation. . . . Scarcity and competition, and cooperation, are two sides of the same evolutionary coin."[14] In other words, anticipation is worth more than extra speed and strength. A prospecting mind must see and "feel." We must size up, anticipate, and estimate the landscape and the best course of action, then learn by experience and adaption. Our prospection allows us to evaluate future possibilities and use them to guide thought and action. The ability to communicate and project with others far into the future and coordinate our actions distinguishes humans from other social animals. Human wisdom is the outcome of human nature, not its origin.

Tomasello and de Waal don't say how our puny primate ancestors outcompeted so many highly specialized herbivores and carnivores. So how did a minor and biologically ill-equipped species rise to such supreme ecological dominance as to reshape the world and evolution itself?

Instead of competing niche for niche with specialized carnivores and herbivores, hominids did just the opposite. They became generalists, hunting a wide variety of animals, from termites to elephants, and collecting and preparing a diversity of plant foods, ranging from fruits and seeds to roots, tubers, and palatable leaves. Tools, fashioned and used adeptly, can dig up roots and tubers more effectively than baboons, pigs, and porcupines, break open tougher nuts than elephants can, reach higher into trees than giraffe do, dig deeper into crevices for grubs and honey than any mongoose or badger, and grind up plant seeds more finely than a bird's gizzard. With their large brains and cooperative learning transmitted among groups and down through generations using language, humans could pass along what they had learned about the behavior and ecology of plants and animals, and so anticipate where, when, and how to harvest them more efficiently than their competitors.

Collectively, these traits made our early ancestors the ultimate multiniche and multi-tasking species, foraging more eclectically and widely than their competitors. In hard times early hominids no doubt reached far beyond the water-restricted range of other large mammals in the same way African hunters still do—by digging down in dry river beds, sucking on watery tubers, and carrying water in ostrich eggs and gourds on foraging trips days away from water sources. With water sources no longer such an ecological barrier, early humans expanded their home range and scouted widely, as I found Maasai scouts doing in searching out pasture when grazing ran short in Amboseli.

Our ancestral success was also spurred by breaking yet another ecological barrier—prey size. The prey size a carnivore can kill is constrained by its own size—and by its muscle power, fangs, and claws. Early hominids broke through these biological limitations by fashioning lethal stabbing and throwing weapons made of wood and stone, and by banding together in hunting parties. Tools more lethal than fangs and claws in the hands of carefully planned and coordinated hunting parties gave early hominids prospects of preying on abundant herds of ungulates as well as megaherbivores like elephants and hippos too hard for the largest carnivores to tackle. The same tools and cooperative hunting skills that expanded hominid prey base in the savannas also helped rebuff attacks from lions, leopards, and hyenas, and often allowed hunters to rob them of their kills.

Ratcheting up hunting skills called for a great deal of cooperation in killing big, fast, and dangerous prey, reducing the risk of death or injury, and sharing the spoils fairly to avoid social disruptions. The link between prey size, hunting band, and food sharing is still found among contemporary hunters. De Waal notes how it takes an entire family group of Lamalera whale hunters to kill and dismember the leviathans, whereas smaller prey like hares are hunted individually. Fairness goes hand in hand with communal sharing and survival. Each member benefits from the larger spoils of collective hunting by monitoring and reinforcing sharing rules that guarantee the individual's own portion and punish cheats who undermine group cohesion.

Though societies have changed since the Pleistocene, present-day hunter-gatherers give us a glimpse of our past. A survey of 339 hunter-gatherer societies by anthropologist Marcus Hamilton and colleagues in 2007 documents a surprising result: our tendency to live in dense populations is not an outcome of modern village and city life.[15] Hunter-gatherers also tend to live in large groups, ranging from a few dozen to several hundred, and for a good reason. For every doubling of population, the home ranges a hunter-gatherer group needs for survival increase by only 70 percent. In other words, large groups afford more benefits to the individual at reduced travel and foraging costs. We shall see this "return to scale," as economists call it, or "the ecology of scale," as I call it, as a recurring theme in the growth of human populations with the emergence first of agriculture, then industry, and now the megacities of the twenty-first century.

Tomasello suggests that two steps led to cooperative hunting: group collaboration in foraging, and the development of shared goals and directed attention. Shared goals led to joint communications such as pointing and gesturing, directed activity, the use of symbols, and monitoring the action of others. The benefits of planning and coordinating the hunt, dividing and sharing the spoils, cooperative child rearing, and a division of labor fostered a group mind. Group mindedness in turn fostered common cultural conventions, norms, and institutions. Individuals could now begin to think beyond themselves to common goals and collective intentions.

I was fortunate growing up to learn bush skills firsthand from African hunters. Nothing matches the chase for its exhilaration and bonding. Although I enjoy an evening out with friends over a fine meal, I've yet to experience again the comradery and thrill of hunting for the pot out with an African hunting band. You feel the tension build as the party picks up fresh spoor, in the hushed whispers about the age of dung piles, where the herd is heading, the number of animals, and what they've been feeding on. Then come the hunched stalk and hand gestures as you size up the prey, close in, and select a target. The final breathless moments of the kill are broken by a sudden rush as the animal falls. Afterward, the hunters gather excitedly around the kill and reenact the hunt before settling down to skinning and dismembering the carcass and lugging it back to camp.

The forest in East Africa is a far harder place to coordinate a hunt than the open savannas. Prey animals, mostly duiker, are small and hidden by dense foliage. Out among the Mbuti pygmies of the Ituri Forest, I am impressed by the tedious effort they put into weaving nets from the inner bark of a climbing vine, *Manniophyton fulvum*. Each net is twenty or so meters long and a meter high. I am awed at how nimbly the hunters navigate the forest and thread nets in a wide arc in the understory. Each hunter guards his section, waiting for the line of women, kids, other relatives, and dogs to drive in duikers with their raucous bellows, barks, and bush thrashing. Each drive lasts an hour or so, followed by three minutes to dismantle the nets, ten to roll them up and lug them to a new site, and another six to reset a fresh line. Hunters rotate their station from center to the outer reaches of the nets and back again to ensure an equal chance at netting prey. Some hunters, women among them, cooperate in weaving nets and dividing the spoils according to their contribution.

I watch from behind the net as a hunter clubs a *moimbo*, a diminutive antelope known as the water chevrotain. Small duikers are cut up and stored in the women's baskets and carried to camp. Later, a large *sindula*, a yellow-backed duiker, is speared as it jumps over the net into the line of waiting hunters. By the time I reach the kill, the men are huddled together in a jovial mood cutting up the sindula and sharing out morsels. Unlike small duikers, which are claimed individually, large animals are shared out among camp members.

The !San hunters of the Kalahari, where wildlife is thinly scattered and more migratory than in the Ituri Forest, track down animals over great distances and can range far from water sources by using ostrich eggs and gourds. New research shows complex reciprocal arrangements among the !San reaching far beyond a simple short-term loan and return of an article, to include complimentary goods and favors reciprocated over many years. Like Maasai social networks reaching across the savannas, !San reciprocity is a latticework of friendships and alliances stretching across the Kalahari.

The many anatomical, physiological, and ecological barriers overcome by our savannas forebears gave them a competitive edge: not just in bipedal walking, but in walking and running great distances; not just in freeing the arms to make tools and throw stones, but in throwing sharpened projectiles with an elastic explosiveness that can kill animals at a distance; not just in bigger brains able to anticipate an animal's movements and track it down, but in minds geared to cooperate in hunting, communicate information, and transmit skills to family and band mates.

In breaking these biological barriers, our early ancestors expanded their food niche and environmental envelope, allowing them to scale the ladder of ecological dominance and widen their range. Breaking the biological constraints holding other species back was a formidable challenge, though, calling for nothing less than uncoupling the very physical design principles that govern the mammalian body. Let me explain why.

Evolutionary biologists long held the view that life-history traits in mammals—birth rate, growth rate, breeding age, and lifespan, for example—are independent of one another, and selected individually by natural selection, like a kid teasing apart pick-up sticks. Darwin saw evolution acting on the whole organism no less than on individual traits. We are not made up of a jumble of body parts—of heart, lungs, gut, liver, kidney, and dozens of other organs each doing its own thing. The delivery of oxygen from the lungs and nutrients from the stomach must be channeled fast enough through the arteries and capillaries to the distant-most cells for them to produce energy adequate to enable the body to function optimally. The amount of oxygen in the blood depends on our lung capacity and breathing rate. The gut must be sufficiently large to supply our daily food needs, yet not so big that it crowds out other organs. All the organs must be tightly packed into the least possible space for our body to work efficiently.

Not much attention had been paid to Darwinian design constraints until half a century after he expressed them, when a British biologist, D'Arcy Thompson, showed how limbs and other anatomical features of animals vary in proportion to one another and with body size.[16] Known as size scaling, or allometry, the beauty of Thompson's idea lay in linking body proportions and shapes to physical laws. An elephant's bough-like limbs are far thicker in proportion to its body weight than the spindly legs of a gazelle. Doubling the size of an animal in height and width increases its weight eight times, calling for stouter limbs to support the load. A one-hundred-ton blue whale buoyed up by water would need limbs so stout on land that it would collapse from its own weight. The limb width, bone and muscle mass, and skin surface area of any mammal are coupled to and constrained by one another and tied to body weight by anatomical and mechanical design principles.

Max Kleiber, a Swiss-born professor at the University of California, Davis, added another dimension to size scaling: physiology.[17] The metabolic rate of mammals varies with body weight in much the same way that the fuel consumption of a car engine varies with its horsepower. The metabolic scaling law, as it is known, is such that for every doubling in body weight, energy production increases by only 75 percent. In other words, larger animals, pound for pound, use less energy than smaller animals.

I added yet another dimension to size scaling laws when my attention was drawn to the links between body size, life history, and ecology.[18] At the time I was working with paleontologist Anna Kay Behrensmeyer of Yale University, now at the Smithsonian Museum, testing whether the animal bones on the surface of Amboseli accurately represent the living population of zebra, wildebeest, buffalo, elephant, and other species. If so, we could infer that the fossil assemblages likely represented the original fauna too. It turned out we could. The bone assemblage on the surface of Amboseli closely tracked the composition of the living large mammals as their relative abundance changed with shifting habitats and climate. In cobbling together findings on dozens of species ranging from a seventy-gram elephant shrew to a three-thousand-kilo elephant, I was astonished to find something far more profound and far-reaching for evolution and ecology: birth rate, growth rate, age at first reproduction, and length of life in mammals all correlate with body weight and one another.

The finding flew in the face of long-held views that life-history traits were selected as independent evolutionary adaptations. I was even more intrigued to discover birth rates scale to body weight in the same way as metabolic rate does to body size, suggesting that the same scaling laws dictate not only our food intake, oxygen consumption, and energy production, but the very pace of life itself. Elephant shrews and elephants have roughly the same number of heartbeats in a lifetime and produce the same number of young, allowing for differences in litter size. Small animals simply run through life faster, like a high-revved two-stroke outboard motor wearing out faster than a four-stroke engine cranking over at half the speed.

Quite why metabolic rate and life-history traits are linked to body size was not fully understood until physicist Geoffrey West and his colleagues at the Santa Fe Institute in New Mexico showed that all animals, from amoebas to blue whales, are built on the same cell design and so need the same amount of oxygen and nutrients to function and power growth and reproduction.[19] The bigger the animal, the more cells it has, meaning the more oxygen and nutrients it needs and the bigger its heart, lungs, and stomach must be to supply them. Yet the length of arteries and capillaries also increases with body size in order to service the distant-most cells and remove waste products. The longer the arteries and capillaries, the greater the frictional resistance in the blood supply and the slower the delivery of oxygen to the cells and the rate of growth, reproduction, and aging. West and his colleagues went on to show that the same scaling laws govern the growth of cities, companies, and economies. I will return to this remarkable finding and its implications for human growth and planetary limits in chapter 21.

The constraints of energy demand and body design posed a formidable barrier to our ancestors in expanding their range and diet as well as combating competitors and predators. Take, for example, the challenge of expanding human brain size from ape size. Our brain accounts for a quarter of all the energy we consume, nearly three times as much as a chimp brain. The expensive brain hypothesis claims we must make up for this large demand in one way or another.[20] Our gut, as in all mammals, uses a lot of energy in digesting food, yet it is far smaller than in similar-sized species. How, then, can we pack in and deliver enough food to power our body, let alone our hungry brain?

Richard Wrangham of Harvard University sees the answer in a richer diet. In *Catching Fire: How Cooking Made Us Human,* Wrangham makes a strong case for the hypothesis that the invention of cooking, rendering meat and vegetable fiber more digestible, allowed humans to forgo a large gut in favor of a large brain.[21] His theory hinges on evidence that fire use emerged when hominid brains began expanding from australopithecine size 2 million years ago. So far, there is no evidence for fire use before 1 million or so years ago.

There is another way to explain what supplied the added energy the enlarged brain of early hominids called for, one that likely evolved long before the invention of cooking. Pounding and pulverizing plants is the equivalent of a cow chewing the cud to break down the thick cellulose walls of plants so that microorganisms in its gut can speed digestion and release nutrients. I long suspected our early ancestors had figured out a mechanical alternative for breaking the cellulose barrier when Mary Leakey pointed out to me what she interpreted as anvils for pounding nuts and grains at Olduvai Gorge, dating from the Paleolithic, a million years or more ago. Was this our early ancestors' equivalent of giant mechanical molars for breaking down coarse fiber and tissue? Recent research confirms that hominins were slicing up meat and pounding vegetables long before blending machines, perhaps as early as 2.5 million years ago. Slicing, dicing, and pulverizing meat and plant leaves, corms, and seeds increases their digestibility by 17 percent, enough to offset the energy demands of our early ancestors with brains half the size of ours.[22]

Our ancestors may also have found yet another way to energize our hungry brain. Biologists took it as given that our metabolic rate—the amount of energy we consume each day—is the same as other great apes of similar size. This is not the case, according to recent experiments by Herman Pontzer of Hunter College and his colleagues.[23] Our metabolism runs faster than that of apes of an equivalent weight—fast enough, in fact, to supply our energy-hungry brain. Our higher metabolism doesn't obviate the need for a richer diet, though: we do, after all, have a small gut. But it does make it plausible that our small gut was the result of bipedalism, dexterous hands, and tools able to pulverize plant and animal tissue. The richer diet was likely enough to supply a hungry brain long before the advent of fire.

A bigger brain, in other words, was no aha! moment, a conceptual leap in our evolutionary story. Rather, it took millions of years of rearranging our body plan to give us a suite of new traits enabling us to break out of our narrow ecological envelope, sever the water-dependent leash restricting our home range, and expand our diet to include larger prey and coarse plants without having to achieve the massive size of the dominant savanna carnivore and herbivore.

Despite its evident advantages, a large brain comes at a cost where evolutionary success weighs most heavily—on reproduction. Primates have a large brain relative to mammals of a similar size. Large-brained primates, humans especially, grow more slowly than similar-sized mammals, give birth more infrequently, and live far longer. A baboon takes seven years to produce its first offspring, gives birth every two years, and lives for thirty-five years or more. The similar-sized, small-brained Thomson's gazelle matures and gives birth in a little over a year and can reproduce twice a year, yet seldom lives longer than ten years. In other words, large-brained primates run through life in the slow lane.

Humans have partially offset the reproductive penalty of a massive brain by saving energy in other ways—in the reduction of our muscle mass and our efficient walking gait, for example. We have given up muscle power for stronger tools and more lethal weapons. On the primate scale of being, our brain, three times larger than that of chimps, means humans should reproduce slower than a chimp's four-year interval. As it is, given a good diet, humans can turn out newborns every eighteen months to two years.

Our accelerated metabolism introduces a fresh biological problem: the combination of a revved-up metabolism and a small gut means big-brained hominids ran out of food faster than other savanna competitors in hard times. In yet another break with our primate body plan, hominids vaulted this barrier by storing extra body fat to tide them over periods of food shortage—an energy storage adaptation akin to the camel's hump in water conservation. This neat adaptation served our ancestors well in surviving food shortages but plays havoc with our modern diets when we binge on rich foods and suffer obesity disorders.

The detour into size scaling and life history throws a different light on the seeming disadvantages of human traits in competition with stronger, faster, better-adapted, and more fecund savanna competitors. Our puny build, low birth rate, and small gut are offset by the advantages of bipedalism, dexterous hands, tool use, a perceptive brain, and our cooperative hunting and gathering abilities. These traits collectively gave hominids the ability to see far, communicate, anticipate shortages, project ahead, plan for crunch times, and secure a larger share of the ecological pie by exploiting many foods and ranging widely.

Looking for a single evolutionary breakthrough to explain our astonishing ecological success—whether a big brain, bipedalism, tool use, throwing ability, fire, language, or culture—reflects our boxed-in science disciplines. As the boundaries between them erode, we are getting a better sense of how breaching barriers and pooling skills through social cooperation have been the hallmarks of our success all along.

So far, I've glossed over the greatest puzzle of all in human evolution: how did our early ancestors break the kinship barrier and form large groups that make us the hyper-cooperative animal we are?

Breaking kinship bonds to scale up to large bands calls for solving another sort of tragedy of the commons. We've seen that large groups are more efficient than small hunting and gathering groups. On the downside, large groups with fewer kinship ties and weaker affiliations are far more susceptible to cheating and social disruption than small, tightly knit kin groups. A hunt leader of an unrelated large group is more likely to hog the spoils than a small family band. Mathematical modelers have grappled with equations to explain how large groups can root out cheats who undermine collective benefits, without much success.

One view of how we overcame the selfish gene and formed larger collaborative groups is the idea that conflict between competing groups reinforces reciprocity and cohesion. Ibn Khaldun, a fourteenth-century Moroccan scholar, referred to the group solidarity and cohesion forged by Bedouin tribes overthrowing town communities as *asabiyyah:* strong reciprocity, in modern parlance. Biologist Peter Turchin, in *War and Peace and War,* elaborating on Khaldun's asabiyyah social glue, argues that wars explain the emergence of group cohesion and the rise and fall of empires such as China, Greece, and Rome.[24] I personally suspect asabiyyah emerged long before the advent of wars, most likely at the very dawn of our species, when group collaboration gave us the edge in competition with large herbivores and formidable predators.

Whatever the cause of strong reciprocity, Ostrom's eight rules for collective action solved the riddle that had thwarted models of kinship theory.[25] As shown in chapter 6, the principles work in managing communal fisheries and grasslands, public freeways and airwaves, suggesting that we are primed to cooperate. We have developed customs, institutions, and social rules that through shaming and punishment root out cheats and curb self-interests from undermining larger, mutual, and deferred benefits.[26] As early as 1759 Adam Smith, founder of modern economic theory, noted in *The Theory of Moral Sentiments,* "How selfish so ever man may be supposed, there are evidently some principles in his nature, which interest him in the fortune of others."[27] Smith showed how our emotions and psychological makeup gear us to adhere to what he called natural moral laws. A stranger passing by is inclined to help you out rather than ignore you if you drop your shopping bag and your groceries spill out. Other strangers would likely tut at the first if he kept on walking by. The Maasai and Malpai ranchers exchanged views on animal husbandry, family, and culture as readily as if they were neighbors, rather than lived half a world apart.

Our natural tendency to cooperate has been confirmed in a wide range of research studies across many disciplines in the last twenty years.[28] Physiological studies show that the hormone oxytocin is released by the brain, giving us that "good feeling" buzz, when we extend a Good Samaritan hand, and it declines when we are deceived. The release of the hormone in mother and infant, and even between dogs and owners staring into each other's eyes, helps create bonds and increase trust, the glue of strong voluntary collaboration. Studies on brain imaging and the new field of neuroeconomics are filling in information on the cognitive details and social behavior girding and reinforcing cooperation. So, for example, an influential study employing the Public Goods game, which uses chips as surrogate money in group exchanges, shows that our willingness to invest for common gains, such as sharing the dividends of spoils among peers, is common across societies; it declines when cheating is detected but increases when players punish cheats.[29]

Summarizing the recent findings, evolutionary psychologist Joshua Greene, in *Moral Tribes,* says that we have inherited an emotional tool kit comprising empathy, decency, gratitude, friendship, vengefulness, honor, loyalty, tribalism, self-consciousness, disgust, indignation, shame, guilt, revenge, and other traits common to all humans.[30] Moral consideration typically extends only to those sharing a common heritage and culture, not to other tribes. Greene goes on to say that we have no ready-made solution to the problem of "Us" getting along with "Them." Our moral compass is designed for tribal compassion and inclusion. The opposite is true in reaching beyond our tribal morality. We are primed to treat others as a threat and look out for our own interests. Different tribes have different moralities geared to their group's self-interests. Greene believes we need a meta-morality to bridge moral divides.

The outcome of our elections in Kenya is still dictated by tribal numbers and alliances five decades after independence, not by policies or creeds. The forty-three tribes created as a nation by artificial colonial boundaries are, nevertheless, beginning to bridge the divides as the young generation merges in schools, business, entertainment, and sport, and shares common values, education, culture, and a national identity. Europe's hundreds of warring tribes at the time of the Roman conquest would have found it inconceivable that they too would one day amalgamate into nations and nations into a European Union sharing common aspirations and values.

As we dig deeper into our human nature, we find precursors of our supposedly unique traits in other species. In *The Age of Empathy: Nature's Lessons for a Kinder Society,* de Waal shows that chimps, and perhaps other higher primates, cooperate with non-kin and engage in elaborate exchanges, including trading and planning. In *Are We Smart Enough to Know How Smart Other Animals Are?* de Waal goes further to say we share most traits with other species: the differences are relative rather than absolute.[31]

Ironically, the sentience biologists stripped from animals and replaced with a pared-down selfish gene version of evolution is under assault and veering toward Darwin's view of a century and a half ago: "The differences in mind between man and the higher animals, great as it is, certainly is one of degree and not of kind."[32] David Sloan Wilson, who has long argued for evolution working at many levels, including selecting for greater cohesion and self-sacrifice among competing groups, notes: "We live in a post-selfish gene age. It is hard to see how it led us down the wrong path."[33]

Degrees can make all the difference, though. Chimps may well settle scores and impose sanctions on others. What they don't do is punish on-lookers who don't punish cheats. Such "indirect punishment," as it is called, is a stepping-stone to our conformity with Ostrom-like rules for the larger common good that propelled us from scattered savanna primate bands to a global society. Shades of difference in individual characteristics matter less than the interlocking suite of characteristics that distinguish our species and allowed us to break the biological bonds and ecological barriers that restrain all other species.

Our unique set of traits answers Shirley's question about baboons: if they are smart social primates that colonized the savannas along with us, what stopped them from traveling further down the path toward us and conquering the world? I have argued in this chapter that our ability to break biological and ecological barriers time and again, and to package a unique combination of traits, enabled us to scale up our social realm from troops to bands to clans, kingdoms, empires, and nations, reshape the world to our own ends, and create a new evolutionary era in which our folly or foresight will decide the future of life.

Is there a Rubicon we had crossed by the Late Pleistocene fifty thousand years ago? In one sense, yes. We have so outperformed, eliminated, or absorbed other would-be large-brained cooperative competitors that the gap between us and other primates is the largest since we diverged from the apes some 7 million years ago. We've reshaped entire animal commu-nities and landscapes, exterminating the big, slow-reproducing species like the mammoth in North America, the wooly rhino in Europe, giant marsupials in Australia, the moa in New Zealand, and the elephant bird in Madagascar. We've used fire to burn woody vegetation, creating open landscapes safer and easier to hunt. Our extermination of species in the course of colonizing the world has shrunk the number and size of large vertebrates everywhere, creating ecological ripples through every continent, biome, and ecosystem to create a new evolutionary era, the Human Age.

For all our transformative power, we didn't reach the top rungs of the ecological dominance ladder until the last fifty thousand years. Until then, hunter-gatherers remained minor league players, measured by their tiny share of the food chain. We were mired in the same Malthusian trap as ever, our numbers held in check by the limited natural productivity of the land and competition with other species.

How, then, did we break the ecological glass ceiling on the next step up the ecological ladder and on to global conquest? How did we become so dominant as to reengineer ecosystems, change our role in nature and our treatment of one another and, ultimately, concern ourselves with conserving other species and saving the planet from our own destruction?[34]

10

Domesticating Nature

If you could go back to the Serengeti one hundred thousand years ago in the Pleistocene Age of Mammals, you would pick out large herds of zebra, wildebeest, and Thomson's gazelle milling around on the open plains, and smaller herds of impala, giraffe, and a cluster of buffalo off in the acacia woodlands. You might see a pride of lions on a kill, a pack of wild dogs jogging off to hunt, or a troop of baboons foraging under the trees. What you would seldom see are humans.

Our ancestors in Africa were still few and scattered in small bands. The biomass of wildlife then would have been about the same as now, around thirty thousand kilograms per square kilometer, capped by the amount of forage produced by the rains. The live weight of carnivores would have logged in at around three hundred kilograms per square kilometer, and baboons, the only primate other than us to venture deep into the savannas, at around five hundred kilograms per square kilometer, less than 0.2 percent of the large-mammal total. Humans would likely have come in at around 0.4 percent of the large-mammal biomass, somewhat higher than baboons, though lower than contemporary hunter-gatherers with their more sophisticated food-gathering tools and skills.

The lack of genetic diversity found among non-African peoples today suggests to geneticists that the populations leaving Africa and colonizing the rest of the world some seventy thousand years ago went through a bottleneck, resulting in a relatively small number of ancestors accounting for the large number of descendants today.[1] From those small bands leaving Africa, we have risen to supreme ecological dominance to the point of commandeering 40 percent of the primary production (annual plant growth) of Earth's land surface. More impressive—or alarming, depending on whether this is seen from our viewpoint or that of other species—we and our domestic animals now account for 85 percent of the Earth's vertebrate biomass.[2] As improbable as this figure may seem, given the world's vast undeveloped regions, such as the tropical rain forests and tundra, the numbers are borne out in Africa, the least developed continent and most abundant in wildlife. Kenya, famous for its teeming herds of wildlife in Mara, Tsavo, Samburu, Nakuru, and Amboseli, has six hundred thousand wild herbivores, less than 2 percent of the 57 million livestock.

Measured on an evolutionary time scale, the rise from our tiny population of a few million at most in the Late Pleistocene to the nearly 8 billion people on Earth today would look like a rocket hurtling into space. How did we ever breach the ecological limitations of a lowly omnivore in the African savannas to dominate every ecosystem, except for the polar latitudes and glaciated mountains?

The home range of modern humans in Africa before the diaspora covered 55 million square kilometers. Our numbers would have grown as our range expanded into Europe, across Asia, and into Australia, an area twice the size of Africa. Despite the lack of genetic diversity, the small diaspora populations spawned a great radiation of distinctive cultures adapted to the new climates, habitats, animals, and plants they encountered: mammoth hunters in the far north, auroch and deer hunters in the south, fishing communities around coasts and lakes, camel and gazelle hunters in the desert. Diet, tool kit, foraging practices, dress, and language began diverging as humans adapted to new environments at a quickening pace.

Starting some eleven thousand years ago, human populations surged from some 2 million to 300 million during the heyday of the Roman Empire around 400 BCE.[3] No other large vertebrate in evolutionary history has climbed so fast or so profoundly altered ecosystems and the physical structure of entire landscapes and continents.

The ecological breakthrough propelling our populations and the ecological makeover of the landscape emerged with farming and herding. Human domestication of other species is not in itself an evolutionary novelty: fungal gardens and aphid farming in social insects originated tens of millions of years earlier. For all their success, the ecological impact of social insects is dwarfed by the human transformation of every biome from desert to tundra.

You get a sense of the transformation in comparing hunter-gatherers, pastoralists, and farmers in our African homelands, where rain still limits plant and animal production. The social and ecological changes are profound. I was first drawn to how few and scattered are hunter-gatherers compared to pastoralists and farmers when out hunting with the Hadzabe in the 1950s. Wiry and seeming to float across the landscape as effortlessly as giraffes, two hunters guided us on a search for a big tusker on a baking hot day. I was too young and fired up by the hunt to realize I was among the last of East Africa's hunter-gatherers. Not until I visited the !Kung in the !Ai !Ai Hills of the Kalahari in the 1970s did I take an interest in the ecology and vanishing lifestyle of hunter-gatherers.

Accompanied by Megan Biesele, who had learned the !Kung click language in studying the role of song and dance in unifying the Ju/'hoansi group, I noted that the Kalahari was drier and the grasses taller than in the savannas of East Africa. The Kalahari's rank grasses are shaped by the single annual rain season in Southern Africa, East Africa's shorter, richer grasses by the double season of the equatorial belt. Wildlife herds are far fewer and more scattered in the Kalahari due to the long dry season, coarser grasses, and paucity of water.

All the hunting skills I learned growing up among Bantu hunters in East Africa paled compared to the prowess of the !Kung. These hunters pick up on the faintest sign of spoor, nibbled leaf, hair, or scent of their scattered quarry; they mime and act out the behavior of their quarry, infer where their prey must feed, water, and rest, and plan how to hunt them down. Out foraging with women, I was astonished by their birdlike vision enabling them to pick out thin tendrils of tubers and their skill in uprooting them with a thin digging stick.

The !Kung's renowned tracking skills have popularized the view that hunter-gatherers are unsurpassed in their knowledge of wildlife. Rich as their knowledge may be to an anthropologist, I was disappointed by their rudimentary grasp of animal behavior compared to pastoralists. As hunters, their insights are largely confined to how to track and kill an animal, just as mine had been on the elephant trail. And no matter how skilled they are in living off wild plants and animals, their populations remain low compared to farmers and herders husbanding their own food supplies.

The pastoralist who spends dawn to dusk herding must know when and where to move his animals, anticipate the behavior of wild herbivores and predators, and know how to outmaneuver them. To get the most out of his herds, he must learn about animal nutrition, milk production, breeding, diseases, the plants to select and avoid in tending his herd, how to choose the hardiest animals and best producers—in short, he must ultimately become as intimate with their behavior as he is with that of his own family and friends. British vets in the 1920s reported that the Maasai exposed their cattle to weak strains of the virulent rinderpest virus to avoid larger fatalities. Out herding with Parashino and Kerenkol, I learned to be wary of buffalo herds to prevent my two cows, Sotwa and Matingab, from contracting East Coast fever, and to keep them away from wildebeest during calving seasons to avoid them contracting malignant catarrh transmitted by the afterbirth of newborns.

Quite how successful pastoralists are ecologically emerged only when I pooled the results of my Amboseli studies in the early 1970s. Until then, the prevailing view was that pastoralism was aberrant, inefficient, and destructive, the epitome of the tragedy of the commons. Ecologist Lee Talbot, working in the Serengeti, had produced figures alleging wildlife to be four to six times more productive than livestock in natural systems— by which he meant national parks. His figures reinforced the view that game farming was more productive and less destructive of the land than livestock. The ecological argument held that wild animals are indigenous and natural, and so highly adapted to the savanna and superior to alien domestic animals. Feeding trials would later show livestock to be as efficient as wildlife in digesting fodder but did little to dampen the hostility of hard-line conservationists toward pastoralists and their livestock.[4]

A feature of the efficiency of pastoralists in the savanna is the far higher density of herders than carnivores. Fewer than fifteen cattle support a milk pastoralist in all but drought years. It takes ten times as many wildebeest to enable a hyena, half the weight of a human, to survive without depleting the herd.

There are of course downsides to managing domestic animals in the savannas. The herd must be corralled at night in a heavily fenced compound, and after a few months the land is stripped of brush, used for fencing, and of grass, foraged and trampled by the animals. The bare earth and rutted trails look as blitzed as a heavily trampled waterhole at the end of the dry season. Range officers cited the denudation around settlements as proof that pastoralists destroy the land like locusts. Stay around long enough after a settlement is abandoned and a different story emerges. I was so intrigued by the rich afterlife of a settlement that I had a student, Andrew Muchiru, study the succession of plant and animal life on abandoned settlements dating back a century.

A few years after a settlement is abandoned and the one-and-a-half-meter mound of dung thins with the rains, winds, and trampling, an ecological transformation takes place. The nitrogen- and phosphorus-rich dung deposits create fertile hotspots and a succession of plants and animals. The pioneer grasses are rich in protein and a magnet for zebra, wildebeest, and gazelle. Over the decades the grasses are edged out by low shrubs that attract rhinos, Grant's gazelle, ostrich, and impala. A half century later, trees grow to supplant shrubs and draw in giraffe and elephant for a century or more.[5]

Flying over savannas, one can see abandoned corrals pocking the landscape like atolls scattered along the Great Barrier Reef. During the rains, thick mats of green fairylike rings of grass erupt on the fertile soil of abandoned corrals. For every new settlement, hundreds of long-abandoned corrals dot the landscape and colonize the bare ground around them. Over the course of a century, the archipelago of ancient corrals creates a shifting mosaic of grasslands, bushlands, and woodlands supporting a profusion of wildlife.

The productivity and impact of herding, though more widespread geographically in East Africa than cultivation, is an ecological pin drop compared to farming. Farmers have modified every facet of ecosystems by moving to the bottom of the food chain and replacing natural vegetation with domestic crops to live directly off the primary productivity of the land, rather than move up to the apex of the food pyramid and live off the pastoralist's proceeds from his herds.

Out among hunter-gatherers, it takes a trained eye to distinguish the human imprint sculpturing the plant and animal community from the natural forces shaping the land. Fires are more frequent, as the coarse grasses are burned earlier to create fresh growth and attract prey. Tree cover is thinner and wildlife more skittish. In the pastoral lands the human imprint is unmistakable in the heavy erosion around settlements. But settlements are few and scattered, giving the illusion of a landscape looking like the classical African savanna of *National Geographic* films. Except, that is, for the livestock mingling with wildlife, which filmmakers screen to portray the savannas as the last of the wilds. Indigenous familiarity or an ecologist's eye in the savannas can detect the human imprint in the patchwork of grasslands, scrub, and woodlands. The footprint of African farmers has, on the other hand, transformed natural ecosystems into intensely humanized landscapes and modified every facet of the terrain, from hydrology to soils, vegetation, and animals, as I found in the Ethiopian Highlands.

I had visited the Ethiopian Highlands several times to set up conservation programs but never had the chance to learn about the ancient farming practices until I joined Walde Tadesse in 2005 to help him plan a workshop on biological and cultural diversity for the Christensen Foundation. Here, in one of the most intensely farmed regions in Africa, I found the same figurative umbilical cord linking culture, society, and livelihood as in the Maasai notion of erematere and osotua.

Walde Tadesse, born in the Gammo Highlands of Ethiopia, earned a doctorate at Oxford and blends a deep understanding of traditional highland farming methods with the cross-cultural sweep and global perspective of a trained cultural anthropologist. I couldn't have had a better companion to introduce me to the country's rich heritage and bat around ideas about culture and biological diversity.

Driving south from Addis, we climb up past the provincial capital of Chenche into the Gammo Highlands, Tadesse's family home. He is choking with emotion as he shows us his seven-hectare field planted with teff, barley, fruit and timber trees, stands of bamboo used for building, and four hundred new apple seedlings. He explains the allocation of land among farmers, and how the community cooperatives help lighten the workload and boost farm production. Each family has several plots spread from the lower valleys to higher elevations, planted with a variety of crops adapted to the altitude and temperature. Above the last of the farms where the trees give way to grassland are the communal grazing pastures. Here youngsters sleeping in well-insulated huts against the biting night cold tend the family herds of cattle, goats, sheep, and horses. We reach the highest peak of thirty-six hundred meters, still heavy with ground mist and frost mid-morning, and look out over densely farmed mountains and valleys.

The steeply folded mountains are Tadesse's ancestral home, a place where customs, livelihood, and culture are deeply intertwined. His world is rich not in biodiversity, for the natural forests have long since been replaced by a patchwork of farms, but in his mind—every plant, animal, person, and inch of the land is an extended member of his family circle. Tadesse explains the elaborate social networks, rituals, and festivals that govern planting and harvesting, coordinate activities, keep labor costs low, boost productivity, and ensure peace and cooperation among farmers, herders, and traders. As with the Maasai, I am struck by how deeply enmeshed his life is in the land and his society, and by how many generations it has taken to achieve such harmony.

Driving south, we descend from the remnant forests of the Gammo Highlands to dry scrub and stony hills looking like the Mukogodo Hills, Shirley's baboon study site in the shadow of Mount Kenya. In stark contrast to the denuded and gullied landscape of Mukogodo, the steep hillsides farmed by the Konso people are chiseled into a series of terraces stacked one above the other like the bleachers at a ball game. The terraces, designed to trap and store rainwater, are built of large rocks needing muscular labor and regular repair to prevent rainstorms washing away the soils and seeds.

Over the centuries the Konso have evolved an elaborate social structure built around collectively farming the arid slopes. A *parka* cooperative is made up of ten or so associates who help in the heavy labor needed on one another's land, most often youths who build and repair the stone terraces, plant crops, and do the weeding and harvesting. A parka can also be made up of women or men who coordinate their activities and hire themselves out.

We stop at Purkuda, one of many villages dotting the ridge tops wending into the distance. The cluster of huts is surrounded by a high stone wall punctured by two narrow gates leading into a warren of narrow alleys. Family compounds flank the alleyways, each with mud-walled huts covered by a thatched roof topped with a large clay pot to ward off rain. Cattle, sheep, and pigs are kept in small stone-walled corrals and fed cereal stalks and fodder. The well-fortified villages are a striking contrast to the scattered individual homes of the Gammo. The zero-grazing system produces more meat and milk than do free-ranging animals, protects the crops from being trampled, and safeguards the livestock from cattle rustlers in the pastoral lowlands. The density of people in the village far exceeds any I have seen in East Africa. One village we visit has twenty-seven hundred people and several *moras,* huts where men meet, talk, and sleep during busy labor periods, and at other times use for public meetings and entertainment.

Farming yields enough food for most families, but some specialize in trading with pastoralists to the south and highland farmers to the north using the *fuldo* system. The fuldo is a network of traders who extend them credit, services, and hospitality on a trust basis. We call on the head fuldo, a craggy old man of great intensity who goes by the name of Usallemeta. He explains how the fuldo system connects traders across the Konso tribe and far beyond. The many local sub-fuldos deal with minor trade disputes and refer serious cases to one of the thirty or so chief fuldos. The head fuldo is chosen by a council of elders. Once the disputants enter his house, the truth must be spoken, as he put it. Ignoring the fuldo's ruling is punishable by excommunication from any community support, even an ember to light a fire. The door is closed, as Usallemeta told me. The outcast becomes a non-person, a pauper. To ask forgiveness he must appease the fuldo with a tuft of grass and may be reinstated if he pays the prescribed fine. The head fuldo alone has the power to make oaths and cast or withdraw ritual curses.

The following morning, we are introduced to another central figure governing Konso society, the *kala*. His compound behind a stone wall topped by juniper posts includes several interlinked houses, among them a ceremonial hut storing his father's mummified body, which will be revered for nine years before burial. The propitiations and fetishes lining the walls and hanging from the ceiling include snake bones and animal skulls. In the compound a baby baboon is chained to a post to deter crop-raiding baboons. The kala has inherited his position from his father and spends his day meeting clan members, dealing with their problems, and mediating disputes. Parka members maintain the kala's field so he can devote his time to the community's welfare.

Although sorghum is the staple in the lower elevations, the Konso rain-harvesting terraces enable them to grow dozens of other crops, ranging in size from trailing lentils to the largest trees. Every species of plant is tended for one purpose or another—grasses for thatching and fodder, herbs and shrubs for the browsing animals and medicines, and trees for firewood, poles, and timber.

The extraordinary diversity of the Konso farming system would be impossible to manage without marshaling human labor and ingenuity to raise productivity above the labors of an individual farmer. The cooperative parka system depends on scheduling the activities of the kala groups through the seasons, governed by ritual ceremonies. Variations or adjustments are made by the fuldos in discussion with the community.

The number of crops planted by Ethiopian Highland farmers is astonishing, ranging from forty in the wetter mountains to twenty in the drier elevations. Ethnobotanist Leah Samberg documented over fifty different crop species in all.[6] The varieties within crops match the diversity of species. The Ethiopian banana (*Ensete ventricosum*) has forty-one named varieties, for example.

The most striking feature of the Konso lands is the remarkable social stratification that reaches from the highest authority to the smallest farmer, and the lattice of kinship and friendship networks connecting the community across the landscape through reciprocal labor and trading systems. How many centuries must it have taken for such a sophisticated agrarian system to evolve and the Konso to reconfigure their landscape to produce such an abundance and variety of crops and animals?

I have learned as much from traditional knowledge and practices as I have from modern science and conservation—I no longer distinguish between them. Knowledge is knowledge, however come by. In his famous physics lectures at the California Institute of Technology, double Nobel laureate Richard Feynman pointed out that science is blind: ideas are measured by the yardstick of observation, not authority.[7] Rocket scientist and peasant farmer are equal when it comes to the test of proof. In traditional societies, what matters is survival and influence. When it comes to the enkingwana sharing of information among the Maasai, you are only as good as the veracity of your information and only trusted if it proves reliable. If so, it will become common knowledge through cultural selection.

Knowledge can also become localized and insulated from changing times and circumstances if too tightly bound culturally. Kenya, despite its poorer husbandry practices and perhaps because of a weaker social stratification, is making the transition from traditional farming practices to a modern market economy far more proficiently than Ethiopia. Greater individualism, social fluidity, and intertribal exchanges make Kenya far more receptive to new ideas, technology, and culture.

One of the long-standing mysteries of our kind is how we domesticated other species in the first place. The Maasai saying "When the belly rumbles, the mind thinks" echoes Ester Boserup's theory that food shortage triggers innovation.[8] A long-held view is that rising human populations and the Pleistocene overkill triggered domestication to offset disappearing wild animals. Evidence from archeological research suggests that the first permanent settlements in the Middle East, perhaps religious ceremonial sites, preceded domestication.[9] The earliest settlements, found in Kharaneh IV in Jordan, date back to 20,000 BCE.[10] Most phytoliths—silica particles in plant tissue—unearthed in archeological digs at the site come from sedges and wetland plants rather than cereals. The large, dense stands of robust grasses would have provided rich beds of edible and readily harvested seeds in the wetlands and seasonally inundated fringes. I can imagine that permanent settlements made it easier for many hands working together to harvest grain, separate the chaff, grind and store the seeds, and guard granaries for the winter months. Dense swards of floodplain sedges and grasses are also easier to protect from wild animals and birds than are scattered fields. It would have taken a small step to boost the yield of favored plants and set the stage for the first small-scale farming systems.

One of the most striking features of domestication is the speed with which it emerged and spread. The first clear evidence of plant domestication in the Near East is emmer wheat, einkorn wheat, peas, and barley, 10,500 to 11,000 years ago; in Asia foxtail millet and rice, 11,000 to 10,000 before present (BP); in Central America bottle gourd, squash, and maize 9,000 to 10,000 BP; in South America beans and cassava, 8,000 to 10,000 BP; and in the Andes potatoes, 10,000 BP. Within 4,000 to 5,000 years of the first domestication, all the staple crops of our modern diet were under harvest. By 7,000 years ago much of Asia's natural vegetation and vast swaths of forest had been cleared for farms.[11]

One can get a sense of the speed and changes triggered by farming from the earliest cultivation of Chogha Golan at the foot of the Zagros Mountains along the Fertile Crescent. Here, archeological digs reveal a continuous 2,200-year period in the evolution of farming societies and crop domestication from wild progenitors of barley, lentils, and peas.[12]

Animal domestication took hold and evolved just as rapidly, as if once our ancestors discovered the skill of molding species to their needs, it spread with all the cultural contagion of toolmaking. With exception of the dog, domesticated from two separate populations of wolves 15,000 to 30,000 years ago, most animals were harnessed for human use within a few thousand years in the Fertile Crescent, Southeast Asia, China, and later the Americas. Sheep were domesticated in Western Asia 10,500 years ago; goats around 10,000 BP; pigs 8,000 BP in China; cattle from an auroch subspecies in Western Asia 9,500 BP; reindeer 5,000 BP in northern Europe; horses 7,500 BP in Kazakhstan; chickens 7,500 BP in several Asian locations; donkeys 5,000 BP, and camels 3,000 to 4,000 ago in the Middle East and Mongolia.[13]

Charles Darwin, in studying animal breeds as evidence for his theory of evolution through selection of natural variation among individuals, was puzzled by another mystery of domestication: the convergent traits among animals of such characteristics as short snout and teeth, floppy ears, black and white coloring, and smaller brains found in many species, including cats, dogs, pigs, and cows.[14] Recent work reported by Tecumseh Fitch confirms the domestication syndrome, as it is called, and shows just how fast it emerges with domestication.[15] Foxes selected solely for tameness in Siberia in the 1950s developed the common domestication traits within ten generations. How could this possibly happen so quickly?

Fitch and his coworkers have come up with an ingenious explanation for the rapid evolution of domestic traits. Neural crest cells arising early in embryonic development migrate to form adrenal glands, parts of the nervous system, pigmentation cells, and portions of the skull, teeth, and ears. Tameness arises from changes in the adrenal glands and sympathetic nervous system responsible for the fight-or-flight response. Young animals yet to develop these responses are naturally tame. If they remain untamed until after the responses emerge, they stay wild. In wolves the window is one and a half months, and in dogs four to ten months. Fitch thinks the late maturation of the neural crest cells in the embryo explains docility. Selection for tameness alone, in other words, creates the shared domestication syndrome.

Whatever the origins of domestication, we can infer from the archeological record and contemporary studies the enormous advantage it gave farmers and herders over hunter-gatherers. Domestication has transformed our wide-ranging daily gathering of hundreds of wild plants yielding a small calorific return into a settled life harvesting a small number of crops rich in calories. The energy-rich domestic crops doubled the birth rate of farmers compared to hunter-gatherers, supported larger and more densely clustered populations, and created larger food surpluses for storage and exchange.

The link between husbandry and culture caught the attention of Russian agronomist Nikolay Vavilov in the 1930s, a time when Soviet state farms were suffering from a dwindling variety of high-yield crops susceptible to drought, cold, and diseases.[16] Worried by famines and mass starvation under Stalin's tyrannical rule, Vavilov set out in search of the ancestors of domestic crops to replenish the lost plant diversity and genetic resilience to disease and environmental stress. Vavilov's travels took him to the Mediterranean, the Middle East, Pamirs, Tajikstan, Mexico, Colombia, and Ethiopia. His seminal discovery was that most of the world's crops originated in a few centers of diversity—in the Fertile Crescent, Karpathians, China, Meso-America, and the Andes—each associated with areas he suspected were plant refuges during the ice ages.

Vavilov also described "biocultures," communities intimately connected socially and ecologically to the land. Biocultures develop and name specialized varieties of crops adapted to specific conditions and refined by husbandry and cultural practices. The feedback between livelihood, culture, and language produces greater crop production, food security, and resilience. Not surprisingly, given his wide travels, Vavilov cited the Ethiopian Highland farmers like the Konso as examples of biocultures that exchange seeds and knowhow to mutual advantage.

The agrarian sharing culture evidently continued to the dawn of industrialization in Europe. Adam Smith, in *The Wealth of Nations*, notes, "Farmers and country gentlemen . . . are generally disposed rather to promote than to obstruct the cultivation and improvement of their neighbours' farms and estates. They have no secrets, such as those of the greater part of manufacturers, but are generally rather fond of communicating to their neighbours, and of extending as far as possible any new practice which they have found to be advantageous."[17]

Social scientists have long argued that the slow tempo of genetic change can't explain the rapid evolution of scattered Pleistocene hunter-gatherer bands into the first complex civilizations in the Middle East in fewer than five thousand years. Several decades would pass after Vavilov starved to death in one of Stalin's prisons, condemned for his Darwinian views, before biologists began to consider culture as an accelerant of evolution. Biologists Peter Richerson and Robert Boyd made a start in their groundbreaking book, *Not by Genes Alone: How Culture Transformed Human Evolution*, published in 2005, in using mathematical models to show how genes and culture can evolve in tandem to speed up evolution.[18] New ideas and innovations, they point out, compete and spread through human populations like genes, yet far faster and more adaptably. The cultural constructs we select and perpetuate vary from words to ideas, ideologies, symbols, livelihoods, social organization, and technology.

Nicolaus Copernicus's and Isaac Newton's theories of planetary motion and gravity, for example, inspired a scientific revolution in showing that our tiny planet rotating around the sun is one of billions of solar systems and galaxies rather than a heavenly creation at the center of the universe. Communism spread rapidly through Asia, Africa, and Latin America after the Second World War, triggering a Cold War between East and West. The Red Cross and Muslim Crescent symbols are universally recognized as giving safe passage to medics treating the victims of war and catastrophes. The Swahili word *safari* spread around the world with the advent of wildlife tourism.

One of the best-documented examples of rapid gene-culture evolution is lactose tolerance in humans. We lose the ability to break down milk by age five. Once we lose the capacity, drinking milk can cause severe diarrhea, cramps, and vomiting. The exceptions are livestock cultures. Since cattle were domesticated ten thousand years ago, five variants of the lactase gene have emerged and spread among African and Eurasian pastoralists and Middle East camel herders.[19]

The speed of lactose spread is no surprise, considering the huge calorific boost it gives pastoral herders switching from a meat to a milk diet. Truly astonishing is the few genetic mutations in humans it took for agrarian civilizations to emerge from hunting and gathering societies compared with the bewildering variety of livelihoods, cultures, languages, and technological innovations.

Ten thousand years of specialized farming practices and technology have shaped our psychology and social affinities as well as our genes and culture. A 2014 study by Thomas Talhelm and colleagues found that farmers of southern China are intensely cooperative in their planting and harvesting of rice, share strong group-oriented norms, are strongly patrilineal, and have a more collective psychological orientation than northern wheat farmers.[20] Wheat farmers live in nuclear households and exhibit more individualistic psychology than rice farmers. The different mindsets are apparent in the way farmers depict themselves relative to their community members. Rice farmers draw circles of similar size of themselves in relation to their neighbors; wheat farmers draw far bigger circles of themselves. Rice farmers are more inclined to reward and less likely to punish friends, favor their own group more than others, and view the world holistically. Wheat farmers are individualistic and analytical in their relationships and worldview. Talhelm and colleagues believe the links between environment, livelihood, cooperation, culture, learning, and psychology have been selected and honed by the different demands of wheat and rice farming.

In a fine summary of gene-culture evolution in his book *The Secret of Our Success,* Joseph Henrich, a former Robert Boyd student, points out that evolutionary biologists like Richard Dawkins and Steven Pinker see human organization, cooperation, and psychology emerging from evolutionary forces shaped by kin selection and reciprocity. Henrich considers both to be important but believes our homo lineage crossed a Rubicon over 2 million years ago with the emergence of *Homo erectus.* From then on culture became the main engine of our evolution. Once cultural adaptations began to outpace genetic changes, the social brain and cooperative capacity of hominids triggered a chain reaction, an autocatalytic process driving ever-faster local cultural adaptations. By observing and copying each other we acquire ideas, beliefs, values, mental models, tastes, and motives that tell us how we should behave in society. Groups learning from each other using cues such as successful hunting, herding, and farming techniques and prestige for the innovators end up sharing similar behaviors, aspirations, expectations, and preferences.

If cultural skills can be accumulated, they can also be lost. Distance and isolation often erode cultural and technological knowhow. Tasmanians had the same complex tool kit as mainland Aborigines when their island separated from mainland Australia twelve thousand years ago, but they gradually lost the use of bone and fishing techniques altogether, reverting to a tool kit as simple as the Neanderthals'.[21]

The domestication of plants and animals has reshaped our relationship with other species and our role in nature as profoundly as it has influenced our relationships within and across societies. We commandeer a lion's share of the energy from sunlight channeled through plants and animals, eliminate and displace competitors, and reshape the environment to suit our crops, animals, and ourselves. For all their success, though, biocultures remained bound to ecological and geographical limitations.

How did we break the constraints of ecology and geography to rise to superdominance at warp speed and create a new evolutionary age, the Anthropocene?

11

Ecological Emancipation

The transition I make each spring from the open expanses of Amboseli, dotted with zebra and wildebeest, to the downtown high-rises, megamalls, and five-lane freeways of San Diego is always jarring. Separated by a flight lasting a day and a half, Amboseli and San Diego testify to our long journey from lives shaped by our Holocene ecosystems to the age of human superdominance in which we are reshaping the world and the very process of evolution.

The explosive impact of our global makeover is hard to grasp in the slowly shifting skyline of San Diego, and harder still to ignore in the accelerating transformation of the Amboseli bushlands into villages and towns spreading along a lattice of new roads. I see the contrasts afresh through Maasai friends awed on their first visit by the bustle and order of American cities, and in the disbelief of American visitors walking among the baboons with Shirley, looking through a window to their primate origins.

The conjunction of the Maasai's small, tightly knit ethnic communities of Amboseli living off the land and globally networked Californians linked instantaneously to a network of friends, chat groups, business associates, newscasts, entertainment videos, restaurant guides, libraries, museums, and millions of videocams beaming live scenes from around the world is illuminating and jolting. My annual spring migration highlights the convergence, collision, jumble, and confusion of cultures over the last two centuries after ten thousand years of spread and differentiation. The outside world advanced slowly on Amboseli prior to the twentieth century, from sporadic trading exchanges with farmers on the slopes of Kilimanjaro to Arab caravans passing through on slave and ivory expeditions. Although the pace quickened with colonialism, Amboseli remained a backwater of Kenya when I embarked on my study of a pastoral community entering the mainstream national economy.

Ibn Khaldun, a widely traveled Moroccan ambassador, judge, and scholar writing in 1377, gave a penetrating insight into how geography shapes people. In the *Muqaddimah: An Introduction to History*, Khaldun details the climate, land, and cultures of the known world.[1] The spherical Earth is half covered by oceans, he begins. The annual migration of the sun, altitude, and climate create distinctive zones stretching from the baking hot tropics to frigid poles that govern the natural productivity of the land and shape human body form and skin color.

In his magnum opus Ibn Khaldun sketched life as a progression of steps: from the earliest development of simple minerals to lowly plants and animals to more complex forms of life and finally to the emergence of our species. Humans are at the apogee of all life, he goes on, distinguished by the capacity of their hands to make tools, their minds to think rationally, and their ability to cooperate in securing food and combating stronger animals. The extent of each community's advancement, culture, and character is shaped by climate and how the people make a living, whether as farmers, herders, fishermen, or traders—here Ibn Khaldun was anticipating the findings of Talhelm and colleagues by seven hundred years.

Following Ibn Khaldun's analysis and Thomas Malthus's portrait of human populations mired in poverty because of constantly outstripping farm production, biologists have assumed that animal numbers are held in check by the limits of food abundance. Ecological studies on thousands of species, from bacteria grown on algal cultures in petri dishes to elephants dying of starvation in Tsavo National Park, have shown plant growth to limit animal numbers. For most ecosystems, energy flow from sunlight and nutrient availability explain the abundance of animal life and the productivity of the land. The best-known and most enduring concept in ecology is the food chain. At the bottom of the food chain lie the plant producers eaten by herbivores, which are in turn eaten by predators at the top of the narrowing pyramid of numbers. Dead animals and plants are decomposed by microorganisms, recycled as nutrients in the soil, and absorbed by plants to start the cycle of life all over again.

No matter how abundant resources are, the productivity of plants and animals is held in check by the least available essential element, most often water, nitrogen, or phosphorus. Known as Liebig's law of the minimum, named after Justus von Liebig in 1840, the concept of critical minima would become a foundation of modern ecology. Another constraint on population is captured by Shelford's law of tolerance: the spread and vigor of a species depends on the ecological envelope of water, temperature, sunlight, and so on it needs to grow, reproduce, spread, and avoid extremes beyond its physical tolerance.

The bottom-up view of plant production regulating animal numbers has been refined as ecologists delved deeper into fluctuations in animal numbers and studied communities as varied as rocky seashores and tropical forests. In one classic study, Robert Paine of the University of Washington found that predators regulated animal numbers and governed the properties of ecosystems.[2] When the predatory common starfish is removed from tidal pools, barnacles and mussels grow in number, and limpets and whelks decline as a result of their increase. Like elephants and livestock in Amboseli, the starfish is a keystone species shaping ecosystems. Paine's experiments in the 1960s added weight to the green world hypothesis, the idea that predators, by reducing herbivore numbers, prevent overgrazing, making the world greener than it would be otherwise.[3]

Another limitation on species size is the range of environments in which they can thrive. Species like baboons, whose body plan allows them to span a wide range of habitats and climates, will adapt more readily to changing conditions.

Recognition that body plan restricts the spread and abundance of various plants, animals, and subsistence societies dependent on domestic herds and crops underlies the notion of *carrying capacity*—the ceiling set on a population by the natural productivity of the land. Carrying capacity often features in debates over how many people the Earth can sustain without risking starvation and population collapse. By breaking one barrier after another—in body plan, kinship ties, Liebig's law of the minimum, and Shelford's law of tolerance—we have raised our carrying capacity and increased our ecological dominance. Dominance in ecological terms refers to how large a portion of the pie one species hogs on a scale of 0 to 1. Ecological dominance is analogous to the Gini index used by economists to measure inequality and the degree of social dominance in a society. In the savannas, few nonhuman species greatly dominate, except in the case of human disturbance—in elephant populations compressed into parks by poachers and settlement, for example.

In the 1970s, when I found Kerenkol dispensing famine relief from surplus American grain stocks, the age-old linkages between rainfall, grass production, and herd and family size were rupturing. I soon realized the futility of monitoring the Amboseli ecosystem minutely and ignoring the impact of larger outside forces. The natural ecological limits and human carrying capacity of the ecosystem were being ruptured by imported energy, nutrients, and water sources. The natural forces governing the Amboseli ecosystem no longer held for humans.

What changed so profoundly in the last few centuries as to propel scattered and highly differentiated subsistence societies of Old World Amboseli into the globally networked manufactured age of New World California?

Measured in evolutionary time, the transition from agrarian to industrial societies happened in a heartbeat. The start of the Neolithic ten thousand years ago saw a steady increase in the world's human population as growth rates doubled from 0.1 percent a year to 0.2 percent by 1500, then doubled again to 0.4 percent from 1700 to 1820 at the start of the industrial revolution. Growth doubled to 1 percent in the early 1900 and yet again to a peak of 1.9 percent by 1950 and on through to the 1970s. Measured in numbers, the world's population will have grown from some 300 million two thousand years ago to 10–12 billion by the end of the twenty-first century.[4]

Adam Smith tabulated figures showing population and wealth to have risen steadily in Europe for a few centuries prior to the 1700s due to investments in labor and capital nudging up food production and surpluses, and the growth of other economic sectors including wool, cotton, manufacturing, finance, commerce, and international trade.[5] Shortly after Smith's treatise *The Wealth of Nations* was published, incomes in Europe accelerated from an average of 100 euros a head each year in the 1700s to 2,500 euros in 2012, despite a surging population. Purchasing power, which had grown slowly in the industrial world between 1700 and 1820, doubled by 1913 and rocketed another sixfold by 2012.[6]

Increased population down the ages has depended on humans exceeding the limits of muscle power, first by using wood and stone tools, then horses and oxen, and finally machines harnessing the energy of water, wind, wood, coal, and oil. Despite the new sources of energy powering the early industrial revolution, Europe by the nineteenth century faced an acute shortage of manure and mulch to keep agricultural output ahead of the growing population. The crisis was alleviated by guano imports from centuries of accumulated droppings in bird and bat colonies in South America, but only for a while: by the turn of the twentieth century the guano ran out and a new crisis loomed.

Based on the work of Justus von Liebig in 1840, agronomists had developed formulae for the optimal ratio of N, P, and K (the three elements critical to plant growth—nitrogen, phosphorus, and potassium) to boost plant yields. The solution to the new fertilizer crisis came from yet another German scientist, Fritz Haber, in the early twentieth century. Nitrogen, which makes up nearly 80 percent of the atmospheric gases, is hard to extract in the form of nitrates plants require for growth. Haber solved the problem by using metal catalysts under high pressure to combine hydrogen and nitrogen to produce nitrates. In 1910, funded by chemical giant BASF, Haber teamed up with Carl Bosch to produce ammonia nitrates on an industrial scale.

The manufacture of fertilizers unleashed a surge in farm production, powered by yet another barrier-breaking innovation, the extraction of energy from fossil fuels to power tractors, combine harvesters, and the transportation necessary for the twentieth-century agroindustry revolution. In *American Agriculture in the Twentieth Century: How It Flourished and What It Cost*, agricultural scholar Bruce Gardner documents the remarkable leap in productivity. In the course of the twentieth century, food production rose sevenfold, farmlands shrank by a quarter, the number of farms dropped from 6.5 million to 2.1 million, and farmers decreased from 65 percent of the population to under 2 percent.[7]

The stunning impact of increasing agricultural output on the American way of life is detailed by Robert Gordon in *The Rise and Fall of American Growth: The U.S. Standard of Living since the Civil War*.[8] Between 1870 and 1940, the average annual output per person rose by a third. The population shifted from 75 percent rural in 1870 to 72 percent urban by

2010. Food costs fell from over half of annual income to 17 percent today. With more food and better medical care, Americans grew taller, infant deaths dropped sevenfold, and contagious diseases such as dysentery, cholera, measles, and polio all but disappeared due to better public health services and inoculation campaigns.

In the late 1800s the family house was lit by candle or kerosene and heated by wood or coal, emitting choking smoke and smut. Men labored long hours on farms, women raised eight or so children—a quarter of whom died in childhood—and spent all day collecting wood and water, cleaning, washing, and preparing food. Without central heating and cooling, winters could be bitterly cold and summers swelteringly hot. With few means of preserving food, the family diet was generally simple and monotonous, and rotten food and contaminated water were constant health hazards. Few people owned more than two changes of clothes or bathed more than twice a week. Families had little time for sport and none for travel. Sons followed their fathers into farming and daughters their mothers in child rearing and domestic chores.

The late 1800s through to the 1930s marked the dawn of modern America, driven by electric lighting and heating, refrigeration, canned goods, and year-round store produce shipped from farms and factories around the world. By the mid-twentieth century the average U.S. home was brightly lit, connected to electricity, gas, telephone, water, and sewer systems, as well as to a world of entertainment through radio and TV. Mechanized transport made commuting to work feasible and the growth of suburban America possible.

Gordon argues that the surge in economic growth in the late nineteenth and early twentieth centuries was driven by a unique combination of innovations and inventions in gas and electric supplies, transportation, medicine, finances, insurance, and communications, as well as by government support for science, public services, and regulation of working conditions. Urban growth offered enormous economies of scale, and the delivery of public services spread from city to rural areas. The low-hanging fruits spurring economic growth have now been plucked, Gordon argues, and the surge in productivity is unlikely to be repeated.

By the end of the twentieth century the United States was a different country and Americans a different people. Liberated from the hard chore of producing their own food, better educated, and enjoying shorter working hours, paid vacations, disposable income, and a new taste for nature, many families could visit national parks and view wildlife from the comfort of a car and experience nature from the ease of public hiking trails.

Although most of the world has shadowed America's agroindustry revolution, hundreds of millions of farmers and herders are still shackled to the land. Those left behind on the fast track of globalization don't share the ebullience of ecologists and tourists over natural landscapes or teeming herds of wildlife. African herders feel much the same about wildebeest wandering out of the Serengeti, munching their grass, and infecting their herds as American ranchers still do about bison wandering out of Yellowstone. The African farmer watching her crops vanish in a whirling flock of Red-Billed Quelea is as hell-bent on eradicating them as the nineteenth-century American farmer was the Rocky Mountain locust and the passenger pigeon. And African farmers and herders have far more reason to feel aggrieved, suffering as they do the double blow of losing their land to parks and their livestock and crops to wildlife.

Americans mourning the passing of the bison forget they are the beneficiaries of the Midwest breadbasket today—among tens of millions more people than the handful of pioneer settlers who struggled to make a living in the prairies after slaughtering the bison and evicting indigenous tribes. Modern farm production helped transform Americans into the wealthiest people on Earth and, paradoxically, led to their first yearnings to restore the bison to its prairie homelands.

In our rise to superdominance we broke anatomical, physiological, and ecological barriers, usurping space and resources from other species and replacing them with our own domesticated plant and animal species. Each step gave us greater freedom from Liebig's law of the minimum and Shelford's law of tolerance. Once we began manufacturing nutrients and engineering our own environments, we gained ecological emancipation from the biological and evolutionary constraints limiting other species to the point of creating a wholly new world, the Human Age. The gains in productivity, efficiency, and the stability of food supplies are hallmarks of our manufacturing success. So too are our narrowed food webs, dependent on fewer intensively managed species, shortened food chains, and the reduction of biological diversity that allows us to slice out for ourselves a bigger share of the ecological pie. All are purposeful modifications of the natural world that explain our rising success, not the missteps of an ecologically aberrant species.⁹ We have done far better than termites in creating our own niche. Where we go wrong is in thinking our conquest of nature is preordained by biblical injunctions to subdue the birds of the air and fishes of the sea or is a natural outcome of our smartness. Yes, we are an astounding success for a feeble ape that made its first appearance on the world stage millions of years ago, but our rise was long and rocky. The fossil record testifies to myriad branches snapping off before the terminal shoots leading to us sprang up. Cooperation and culture, not intelligence, made the difference. Other branches of our ancestry, smart but perhaps less able to collaborate in large non-kin groups—the Neanderthals and Denisovans among them—vanished.

Ecosystem dominance took an inordinate degree of cooperation, as we've seen in the case of the Maasai and Konso. Our dominance of the savannas and Ethiopian Highlands took thousands of years from the time the first farmers fanned out from colonizing centers of cultivation, cleared forests and bushlands, replaced them with crops and livestock, and guarded them against wild animals and hostile neighbors. Individual self-interest and poor husbandry practices caused countless failures before biocultures emerged able to manage the land sustainably and boost productivity.

For all our ecological success, there is no cause for celebration among the poor still feeling the pinch of starvation, or among the richer nations suffering the lifestyle diseases of obesity, heart disease, asthma, and sedentary ailments as well as the particulate matter, pollutants, and toxins pervading our new global nest.¹⁰ The by-products of our manufacturing age and hubris of our ecological emancipation are eroding the natural capacity of ecosystems to decompose and recycle nutrients, retain water, regulate climate, and dampen environmental stress. We are still too mesmerized by material gains to take stock of the cost to our own health and well-being, much less to the future health and resilience of our planet.

We have no evolutionary precedent to draw on in curbing our destructive behavior. We are on our own, an evolutionary novelty not only in our unrivaled success in dominating and reshaping ecosystems and continents, but also in our knowledge about the world and our newfound sensibilities about nature. We have, in short, become the sole superdominant species.

Our immense global impact forecloses the option of escaping our dirty nest by moving elsewhere. We have plucked the low-hanging ecological and economic fruits, kicked reparation costs down the line, and now face narrowing choices and tougher measures if we are to better our lot and repair our self-inflicted damage. Messing up the global commons is not like destroying the forests and pastures of ancient Greece. The raft of unintended and unexpected consequences is so vast, remote, and complex that fouling up the planet is truly inviting tragedy for all.

The changes in our social lives have been no less monumental in the course of our ecological emancipation. Our growing populations and concentration in urban centers have broken down social barriers and created larger-scale societies, a concentration of power, and the rise of political systems that reach across vast geographic regions and tribal boundaries. Francis Fukuyama, in *The Origins of Political Order: From Prehuman Times to the French Revolution,* notes that big advances have been made over the last forty years in explaining the rise of political power and institutions.[11] Contrary to views of philosophers such as Thomas Hobbes, who saw human sociality as a tabula rasa shaped by political order, political scientists see recurring patterns of behavior down the centuries that reflect our deep-seated primate ancestry overlaid by our social propensity and cultural evolution. As populations grew and small-scale bands, clans, and tribes amalgamated into kingdoms, empires, and modern states, they set up a recurring tension between individual and nepotistic interests on the one hand, and on the other the collective benefits of larger, denser populations: greater productivity, security, public services, and innovation. In an interesting twist, Fukuyama sees modern democracy as restoring the glue that bound small-scale communities in the transition to large, complex societies.

Ronald Wright sees civilization—large, complex societies based on domestication of plants, animals, and our own human nature—as an experiment in living within the means of food supply and environment.[12] The subtext is that for complex, dense societies to survive and thrive, they must develop refined husbandry skills, ensure environmental security for their citizenry, and develop rules and institutions for managing the commons. Wright sees civilization as a naturalistic experiment run in parallel in the Americas, Eurasia, and China. He agrees with Joseph Tainter's view in *The Collapse of Complex Societies* that three factors brought down past civilizations: the overuse of resources, cultural inflexibility, and strongly hierarchical and repressive societies.[13]

Ibn Khaldun in his fourteenth-century treatise was remarkably astute in recognizing that altitude, climate, and the annual migration of the sun create distinctive biogeographic zones that shape the natural productivity of the land, and that humans are distinctive in their capacity to make tools, think rationally, and cooperate in bettering their lives and combating competitors. He would be astonished to see how far we have come in escaping the biological and geographic forces he saw as shaping livelihoods and cultures, and by how technology has changed the course of human evolution.[14]

From Khaldun's time onward, guns, germs, and steel played a growing role in the rise and fall of nations, as Jared Diamond argues.[15] With the advent of industrialization, the institutions of modernity and the performance of governments take center stage in the progress of nations.[16] Now, in an age of increased globalization, the degree of our integration into the world's marketplace of goods, services, ideas, innovations, technology, and communications counts even more.

What is the global age, the new era of our own making? How do technology and globalization change us as individuals, as societies, as nations, and how do they affect our role in nature and in shaping the future of our planet?

With my father at my first and last kill (*above*). My passion in life switched from killing to saving wildlife, including a cheetah orphan (*right*).

In 1967 I pitched my tent in a quiet grove to feel the pulse of the Amboseli. Within eighteen months, the sonorous *boop-boop* of the Verreaux's Eagle-Owl at night gave way to the monotonous *tap, tap, tap* of Cardinal Woodpeckers as the woodlands died.

In 1883 explorer Joseph Thomson wondered how such enormous numbers of large game could live in Amboseli.

The more baffling enigma to me was how so many Maasai livestock could coexist with wildlife.

I learned more about every nuance of the savannas from the Maasai in one season than I did from years of research.

Kerenkol wedding
his first wife (*above*).
His son Josh forty
years later in Aspen,
Colorado, where he
loved the ski slopes
(photo courtesy
Jordan Curet) (*right*).
Despite being a war-
rior leader, Kerenkol
saw the Maasai way
of life eroding and
prepared his children
for the new world.

Modern sensibilities help conserve vignettes of culture and nature but change them in the process. Venice, propped up by tourist revenues, risks losing its Venetian soul (*above*). Serengeti migration has become a tourist mecca, losing its wild appeal (photo by James Jiao / Shutterstock.com) (*below*).

Deep-draft cruise ships and sea-level rise are eroding the foundations of the ancient mercantile city of Venice.

Should Adam Lowe reproduce the Veronese masterpiece ravaged and darkened by time or restore the painting to its original vibrant hues? *The Wedding at Cana* epitomizes the dilemma inherent in conserving and restoring cultural heritage and natural ecosystems.

The meeting of two iconic cultures—the Maasai, among the oldest cattle herders on Earth, from the continent of our birth, and the recent Malpai ranchers, settlers of the last continent colonized in our global diaspora. Warner Glenn with Dennis Sonkoi, Yusuf Petenya, and Joseph Miaron.

We head deeper into Malpai Ranch in Arizona on the lookout for cattle and wildlife. Yusuf asks me why the herders have abandoned their cows to lions.

Shirley Strum showed baboons to be humanlike in many ways, yet they took a different path than humans in colonizing the savannas.

Olorgesailie in Kenya's Rift Valley captures the emergence of modern humans as they began breaking ecological barriers and starting on the long journey to global conquest. Here a nine-hundred-thousand-year-old butchery site of an extinct hippo has stone tools scattered around.

American homesteaders abandoned their farms during the Dust Bowl in the 1930s (photo by Everett Historical / Shutterstock.com) (*top*). Today thousands of irrigation booms fed by giant pumps drawing water from deep aquifers have turned the plains into an American breadbasket and cattle feedlot (photo by Alex Pix / Shutterstock.com) (*bottom*).

Small-scale homesteads (*top*) have been replaced by large agroindustrial farms and cattle feedlots (*bottom*).

One way or another, cities will define our future more than any other of our creations, whether the dense slums of Kibera in Nairobi where the poor congregate (photo by John Wollwerth / Shutterstock.com) (*top*), or San Diego, a leading IT and green economy hub in the United States (photo by Jerry U / Shutterstock.com) (*bottom*).

Seen by billions around the world, the *Apollo* image of Earth took us out of our cocoon and showed us how unique and tenuous is the tiny blue sphere supporting life in our solar system (*Apollo* mission courtesy of NASA).

PART

II

The Human Age

12

The Global Express

In his famous thought experiment on relativity, Albert Einstein imagined a passenger on a speeding train and an observer on the platform. To the passenger, the train is static and the platform rushes by. To the observer, the platform is static and the train rushes by. The early and late passengers boarding the global express are Einstein analogues. To Californians at the front end of globalization, houses, stores, roads, utilities, work, dress, and food have outwardly changed little in decades. To Amboseli Maasai newly boarding the global express, the world is changing in the blink of an eye, the gap with the economies of the developed world closing in half a generation.

To the tourist, Amboseli looks much as it did to Joseph Thomson, the first European explorer to visit the area in the late nineteenth century. Zebra and wildebeest still head off toward the Chyulu Hills at first rains and trail back to the permanent springs at the foot of Kilimanjaro when the grass withers. Maasai elders hew to their age-old traditions, sitting under a tree, talking of bygone days, and playing *bao*, a board game handed down through hundreds of generations. And yet each year I see the last vestiges of the Neolithic Age being chipped away as seasonal corrals become permanent homes and villages.

When I set up camp in 1967 in a small tongue of acacia woodlands jutting into the northern plains of Amboseli, the daily traffic jams, blare of horns, and crowded streets of Nairobi 250 kilometers away were replaced by the quiet shuffle of wildebeest and zebra hooves kicking up puffs of dust with each footfall. Rain was the measure of all things. God, *Enkai*, delivered the rains that marked the seasons that regulated the numbers and movement of wildlife, the locale of Maasai corrals, milk production, and family well-being.

Old man Lakato, sitting outside his settlement nestled in a grove of flat-topped *Acacia tortilis* trees at the foot of Il Marishari Hill when I first met him, spoke of past and present as a continuum, an unbroken cycle of daily herding punctuated by ceremonies marking his passage from birth to old age. His social circle, a network of relatives, friends, clan, and section, made up the loose federation of Maasai who had moved from the Nile Valley into Kenya and Tanzania several hundred years earlier, conquered resident pastoral tribes like the Ilogolala, and colonized lands the size of Arizona. Reed thin with receding hair and a grizzled face, Lakato would squat on a three-legged stool gently waving his fly whisk, surrounded by youngsters eager to hear his tales.

Listening to Lakato I was struck even then by the differences between our two worlds: his one of family, stock associates, and friends scattered across his annual foraging grounds, mine a network of anonymous intermediaries reaching back to remote sources of food, news, transport, knowledge, and entertainment, all connected by invisible airwaves, electricity, transport, and a money trail.

Little had changed for Lakato during his lifetime. He still held council with elders under a majestic *Acacia tortilis* tree reputedly old when his grandfather was young. Just as he had been inducted into Maasai customs and culture by his father's generation, so his many children looked to him for the direction and skills needed to raise a herd and a family. Being a Maasai was about knowing your place in society, giving and receiving respect, and earning the esteem, trust, and support that every elder cherishes and depends on for survival and success.

By the late 1960s a small advance guard of Maasai had begun to break with their traditions and move to the outer world, Simon Salash among them. He had made it through primary school, found a job, and was living in Loitokitok fifty kilometers away. Looking back at Amboseli from the outside, Salash told me he feared the Maasai of his childhood would be swept aside by an alien world and strangers eyeing Maasai lands. Even he could not have foreseen how fast highways, trucks, farms, traders, towns, stores, bars, schools, clinics, lodges, and tourists would invade Amboseli and transform Maasai lives and culture.

A half century on, I regularly pass by the scar of Lakato's settlement. Every sign of his existence has returned to earth—his body, loin cloth, sandals, belt, gourds, dung huts, livestock, sleeping platforms, and thorn enclosures. All that remains are a few scattered beads and a rich patch of manyatta grass where his corral once stood. Close by is a forty-bed ecotourism lodge run by Cheli and Peacock Safaris on land leased from Kerenkol, whom I had first met in 1969 when he was the laigwenani, the spokesman for his warrior age group. The lodge is a measure of how well he's done in Kenya's emerging economy. His first venture beyond his customary herding was to buy cattle across the border in Tanzania and sell them at twice the price to backstreet butchers near Nairobi. Next came a small farm tended by seasonal laborers; a truck he rented out to farmers to haul produce to market; a taxi business; and finally the tourist lodge, which paid for his sprawling modern home housing four wives and twenty kids.

Kerenkol too has passed, but not before seeing his son Josh complete the transition from Lakato's cattle world to global voyager. Graduating from high school up-country, Josh took a course in hotel management in Switzerland, a degree in hospitality in the United States, and found a job at a ski resort in Aspen, Colorado. Unlike his father, Josh doesn't measure his success in wives and children. Aged thirty-five, he is unmarried and has other plans. He will fit a family into his career, he insists, marry an educated, working woman and have two kids free to choose their own careers. Josh shares my passion for wildlife and his father's for cattle. He leapt at the chance to join the board of the African Conservation Centre in the United States and promote the community-based conservation his father helped pioneer.

Whereas in the 1960s Amboseli was being slowly invaded from the fringes by development, the changes today are taking place in every home. Virtually all families send their kids to school, hoping they will make it as a pilot, nurse, accountant, lawyer, or doctor—any job paying well enough to build and furnish a brick home with modern comforts. Few families move seasonally any longer, choosing instead to settle near permanent water, schools, stores, and social amenities. Dirt tracks and power lines meander across the bush, linking new villages and towns. Few young Maasai now know what erematere means, and most see conservation as protecting wildlife or, among the widely traveled, biodiversity. As in the West, biodiversity is such a novel notion that Swahili, like English, has adopted a new compound noun for it, *bioanwani*.

A large family herd is still the dream of the Maasai, but it is reality for few. Herders with enough animals to feed a family are Kenya's version of American cattle barons: the wealthy and politically well connected who live in Nairobi, send their kids to the best schools, buy land and livestock, and hire herders to manage them. The hired hands know little about live-stock and nothing about combating droughts, lions, and hyenas. For a while, until conservation organizations began hiring Maasai youngsters to stop the practice, herders dealt with stock-killing predators by lacing a carcass with a lethal dose of insecticide, sending Kenya's vulture population into a nosedive as collateral victims.

In the struggle to do well in the new cash economy, herders raise three times as many sheep and goats as cattle to pay for food, school fees, and perhaps a Chinese motorbike. Most Maasai are on the lookout for jobs and many have become farmers, a life they once disparaged. The fertile areas of Maasailand are being subdivided and sold off to immigrant settlers. Rivers, swamps, and forests once offering pastoralists refuge from drought are being drained or hacked down for irrigation, horticulture, and urban development. Pastures are degrading with the permanent grazing around settlements, reducing the carrying capacity of the land for livestock and wildlife.

David Maitumo is typical of the Maasai moving into the wage economy. Quiet and more at home in Amboseli than town, David became my research assistant after leaving high school in 1977. He soon learned to monitor and record plants, animals, livestock, and settlements, join me on aerial counts, and run the Amboseli Conservation Program in my absence. Tak-ing a cue from his uncle, Kerenkol, David used his salary to start a three-hectare farm, dig irrigation channels from the nearby Enkumi River, and grow onions and tomatoes to sell to wholesalers in the cities. He later added a fruit orchard to diversify his farm economy and bought a petrol-driven mill to grind and sell his maize crop. His brother manages his milking cows and commercial beef herd in his absence.

David's cattle are no longer tied to the seasonal rains and pastures. In tough times he moves his Amboseli herd as far as Nairobi, where he buys grazing rights from farmers, paid for in digital cash transacted by cell phone. From his four-bedroom stone house replete with tiled floors and ceilings, running water and electricity, and a large TV, his wife Joyce sells Maasai bead wear to tourists and chairs a women's group developing a traditional village to accommodate home-stay tourists keen to experience the Maasai life. Unlike Kerenkol, David stopped at four kids and put them all through college. His second son, Sakimba, joined my Amboseli program and researched Maasai views of the changes in Amboseli for his MSc degree. He has gone on to a doctoral program at the University of Lyon in France, studying the Maasai erematere system by interviewing the last of the elders who practiced the method in the 1970s, with a view to restoring their seasonal grazing practices on degraded lands.

Amboseli proved to be the microcosm of the changing world I visualized it would be in 1970. With few exceptions, the Maasai boarding the global express are on a fast track, leapfrogging over old technologies and infrastructure to embrace the IT age with alacrity and astonishing ease. Twenty years ago, Kenya had fewer than half a million landlines connecting wealthier urban Kenyans through a tangle of wires with more splices than a silent movie film reel. Giraffe so regularly snapped our home telephone wires that I bought a long ladder to fix the lines myself rather than wait days for the phone company. Today, Kenya has 35 million users in a population of 45 million, connected wirelessly by $50 cell phones, with rates at a tenth of the U.S. talk-time. Kenya is leading the digital money revolution. Electronic banking and money transfers account for nearly a third of all transactions. Tens of thousands are being lifted out of poverty by ready access to credit and loans to start new businesses, and through the microenterprises and cooperatives that are bringing a growing number of women into the workplace.[1]

Quite how radically cell phones have changed Maasai lives first struck me as I chatted at a Nairobi coffee shop with John Marinka and Daniel Leturesh, chairman of the Ololorashi Group Ranch surrounding Amboseli. Leturesh took a call on his cell phone and I could hear an agitated voice breaking through the restaurant chatter: "A herder fleeing rangers driving his cattle out of the national park ran into a lion and was badly mauled. His friends are armed with spears and heading for the park to kill the lion. You've got to stop them before the rangers confront them." After a few minutes of back-and-forth, Leturesh and Marinka stood down the warriors as adroitly as if they were on site. Halfway around the world Josh Kirinkol uses a cell phone at his new home in Aspen, Colorado, to stay in daily touch with herders out with his cattle, wanting to know every detail about how his animals are doing and listen to the clink of cowbells.

The new age of global connectedness has changed my world of conservation too. In 2012, up on the Tibetan Plateau, naturalist George Schaller, Lu Zhi, director of San Shui Center in Beijing, and I were looking at the overgrazed slopes and eroding topsoil silting up the Yangtze and Yellow Rivers. The Chinese government was paying $2 billion a year on reforestation and settlement schemes to reduce the erosion. Lu had downloaded images of the worst-hit areas on her iPad to compare past and present satellite images. I was puzzled by scattered remnants of juniper trees on heavily eroded slopes. When were the slopes last forested? I ask Lu. She put out an information search on her iPad and moments later had a detailed answer of the deforestation history.

The twenty-first-century world, only two decades old, is already far smaller socially and more interconnected than in the closing years of the twentieth century. VCAMs at lodges and waterholes across Africa beam close-up scenes of wildlife and elusive animals around the world via You-Tube. Camera traps deployed in the Serengeti detect and map animals ranging in size from rodents to elephants by cloud-sourcing. In Kruger National Park plate-sized drones cruise overhead counting animals, mapping vegetation, and scouting out poachers.

Like Kerenkol, I have changed and adapted. I'm not hell-bent on hunting big tuskers in the remote corners of southern Tanzania as I was as a youngster. On the contrary, I do my best to save elephants from a resurgent illegal ivory market. My annual migration to San Diego with Shirley is a voyage across space and through time, from the starting point of globalization in the African savannas to the leading edge in California.

From my small office shack in the paddock of our Encinitas house, I conjure up the California known to the first human colonizers who crossed the Bering Strait from Russia perhaps twenty thousand years ago and, in a larger wave, five thousand years later as the glaciers receded at the end of the Ice Age. California then would have been uncannily familiar to my Amboseli eye. A hundred miles north of San Diego at the Los Angeles

Natural History Museum is a display of African lookalikes—elephant, rhino, lion, hyena, cheetah, and buffalo—all trapped in the La Brea tar pits during the Late Pleistocene. By ten thousand years ago all had vanished from North America, victims, perhaps, of prodigiously efficient Clovis hunters who knapped the most elegant and lethal stone spears ever manufactured until the development of metallurgy.

When the Spaniard Juan Rodríguez Cabrillo sailed up the Pacific coast from Mexico in 1542 to reach present-day San Diego, he found a vast wildland sparsely occupied by the Kumeyaay Indians, who foraged the coastal plain in winter and the Cuyamaca Mountains in spring and summer. After Cabrillo's exploratory voyage Spanish priests arrived and established a string of missions along the California coast to convert the Indians to the Christian god and "civilize" the natives. The wave of American trappers, traders, and miners arriving in the nineteenth century caused a second wave of animal extinctions in California and across the Interior West. Bison, wolf, bear, coyote, cougar, beaver, and turkeys were gunned down, trapped, or poisoned for commercial hide and meat markets, killed as vermin, and shot for the pot and sport. Forests, prairies, and wetlands were felled, plowed, and drained for ranches, farms, and settlements.

The view from my writing shack on the edge of the Escondido Creek looks across a floodplain choked with riparian brush, giving wildlife in the sagebrush hills a corridor to the San Elijo Lagoon five kilometers downstream where the busy I-5 freeway reverberates in an endless stream of traffic. Wildlife has encroached on San Diego over the last few decades despite suburban sprawl. Deer, coyotes, mountain lions, and bobcats have filtered into the riverine woodlands and brushed canyons. North County has far fewer native species than in Kumeyaay times, though a far greater diversity of species overall—if exotics are counted. The new immigrants include thousands of garden plants, farm crops, zoo animals, exotic trees from around the world, domestic animals of all sorts, and unwelcome invasive species including rats, sparrows, and crows.

In the newly humanized landscapes the links between rainfall, temperature, and nutrients that once governed ecology have weakened. Many biologists still talk of the "ecological lie" of growing crops and raising animals outside their natural limits set by Liebig's law of the minimum and Shelford's law of tolerance. They ignore the profitability of pumping water to grow crops year-round in the sunny southern part of the state and heating greenhouses to produce winter crops in the cooler northern part. When humans rejigger climate, landscape, ecology, food chains, and supply lines, the engineered realm can be far more productive than natural ecosystems.

California's Central and Imperial Valleys generate billions of dollars' worth of produce in lands naturally no more productive than the African savannas. Year-to-year rainfall counts for little when fossil fuels pump water through huge aquifers from humid northern California and from groundwater tables deep underground, creating an oasis in the desert. Where climate and ecology will figure is in whether the productivity of those valleys is sustainable in the long term, and how much the taxpayer is willing to fork out to insulate the oases from climate change. The Salton Sea, once fed by a channel from the Colorado River and a playground for Hollywood stars and business magnates in the 1930s, ran dry when farmers, corporations, and city homeowners won out in claiming a larger share of the waters.

For the first-time visitor from rural Kenya, California is a bewildering and daunting landscape, as the African bush is for the first-time American hiker. For the pastoralist raised on the open plains and used to the slow tempo of the seasons, the scurry of people in downtown Los Angeles moving as frenetically as foraging ants, with no sign of fields and livestock in sight, is baffling and inexplicable. Our modern human-built world, as historian and sociologist Thomas Hughes calls it, has changed our culture and thinking beyond the ken of pre-industrial societies.[2]

The paradox of globalization is that the Brownian movement of Los Angeles, inchoate as it may seem to the alien eye, works with astounding efficiency and effectiveness. Adam Smith's invisible hand, guided by city planning ordinances, zoning laws, civic codes, regulations, standards, licensing, protocols, and social norms, makes order out of chaos for the city's residents, businesses, and commuters.

Among the mottled views of globalization, Manfred Steger sees it as the expansion and intensification of social relations on a worldwide scale.[3] The term stems from the efforts of the United States, Great Britain, and their allies to reconstruct the shattered economies and infrastructure of Europe after the Second World War. The Bretton Woods conference of forty-four nations, held in New Hampshire in July 1944, laid out a framework for international commerce and cooperation aimed at boosting economic growth through government and private-sector investments, stimulated by new trade agreements and advances in science, technology, communications, and transportation.

The assumption that free trade stimulates economic growth rests on the theory of comparative advantage first proposed by the English political economist David Ricardo in the eighteenth century.[4] Ricardo stated that two countries should trade if each was relatively better at producing certain goods—say, England's wheat and Portugal's wine. Each country stands to gain from cheaper goods through trade, boosting the economy of both. A key assumption of Ricardo's theory, expanded by Swedish economists Eli Heckscher and Bertil Ohlin, hinges on the elements of production being equally available in every trading country, arising solely from the capital, labor, and natural resources at its disposal.[5] Under the Bretton Woods agreement, the benefits of global trade and economic growth were predicated on fiscal and political liberalization and assistance from the World Bank and the International Monetary Fund to less developed countries to speed investment and development. A host of new global institutions emerged to foster human and economic development, including the United Nations General Assembly, Food and Agricultural Organization, World Health Organization, Human Rights Commission, and International Labor Organization, together with transnational agencies regulating and governing global aviation, shipping, communications, and the environment.

Despite the discontents with the Bretton Woods vision of globalization in many developing countries, economists generally regard the impact of trade liberalization on world economic growth as a resounding success.[6] Over the half century following World War II, the volume of trade rose from $57 billion to $6 trillion as tariffs worldwide fell from 40 percent to 6 percent.[7] Although merchandise trade has since leveled off, digital and information exchanges with virtually no barriers are expanding many times faster than global economic output, inducing a wave of productivity and efficiency.

Incomes and human development indicators—such as infant survival rates, levels of literacy and health, and the quality of air and water—used by the United Nations as benchmarks of development have risen with global market expansion and economic integration. Food shortages have shrunk, family size and population growth rates have fallen, disposable income and leisure time have risen. Most people today are networked around the world by radio, TV, cell phone, and computers. The global bazaar of goods and services has reduced the scarcity of local resources, the role of seasonality in food supplies, and high labor costs, slashing the retail price of consumer goods through lower import taxes, economies of scale, and supply chain efficiencies.

The dizzying speed of globalization over the past two decades caught even seasoned political commentator and *New York Times* journalist Thomas Friedman off guard. In his best-selling book on globalization, *The World Is Flat*, Friedman considers the flattening of barriers in the late twentieth century to be the final phase of globalization, which began in the first phase when Columbus discovered the New World for Spain in 1492, expanded in the second phase with industrialization in the 1800s, and quickened in the present phase with the spread of modern communications and transport, creating a global marketplace.[8]

Other commentators see globalization starting far earlier than Friedman does. Economist Andre Gunder Frank dates globalization to the rise of trade links between the Sumer and Indus Valley civilizations five thousand years ago.[9] The more recently you start the clock ticking, the more globalization appears as a modern phoenix; the further back you start, the slower it emerges. Looking back on the growth of population stimulated by globalization, you can see the hockey-stick inflection of human population growth at the dawn of industrialization. Earlier still, populations surged in the Neolithic and spread with the advent of farming and herding. Earlier yet, populations rose steadily in the Late Pleistocene with the development of sophisticated stone tools and specialized hunting and gathering techniques. Earlier yet, populations rose with the diaspora of modern humans out of Africa and across Eurasia seventy thousand years ago.

I see the story of globalization starting far longer ago. Two hundred meters from my house looking into Nairobi National Park and across to the city skyscrapers beyond, the first steps to globalization are indicated by the obsidian and chert tools dating back three hundred thousand years or so washing out of a small erosion terrace spilling into the Mbagathi River. Alison Brook, Rick Potts, and colleagues have deduced that the obsidian and chert used in fashioning the stone tools at Olorgesailie were transported from the same source, at least two days' walk away.[10]

Olorgesailie marks our species' first tentative steps beyond ecosystem boundaries some half a million years ago, possibly much earlier. Those small steps led eventually to a steady expansion and colonization of new lands and continents, and to innovations in cooperation and tool use that launched our global domination. At the dawn of the twenty-first century, our global voyage and our world domination and fusion of cultures are almost complete. Genetic mutations, once the pacemaker of evolution, have been surpassed by cultural evolution, and cultural evolution is accelerating with technological innovations that have disrupted and flattened the world even as Thomas Friedman knew it a few years earlier.[11] Why is globalization, our human-built world, so alluring, inevitable, and disruptive?

13

Converging Worlds

The Hadzabe hunter-gatherers who guided my father on an ele-phant hunt in 1958 showed no interest in our safari gear, other than the odd item they thought useful—an unbreakable aluminum water bottle, tin cups, and my father's razor-sharp kukri from his army days with the Gurkha troops in India. His proudest possession, a new safari truck, was of no use if we had to call on the Hadzabe to track animals through the thorn thickets, down gullies, and over rocky hills. Strong legs and a good pair of eyes were far better than a car on a hunt.

The life of the Hadzabe lay as clear as a footpath in following their fa-thers' footsteps as hunters and their mothers' as foragers and childbearers. Recent evidence points to how ancient and unchanged is the African hunter-gatherer way of life now vanishing among !Khoisan click-language speakers of East and Southern Africa. Excavations in the Border Cave in the Lebombo Mountains of South Africa near the Swaziland border have unearthed carved and decorated tools, arrowheads, digging sticks, and ostrich beads forty thousand years old identical to those of the !Kung today.[1] No other culture has survived so long unchanged.

The !Khoisan hunters persisted across Eastern Africa until waves of immigrants—pastoralists entering from the north thirty-five hundred years ago, and Bantu farmers from West Africa twenty-two hundred years ago— pushed the hunter-gatherers southward and into marginal areas. The Had-zabe in the dry bush country south of Ngorongoro Crater in Tanzania and !San in the Kalahari of Southern Africa are the last of the !Khoisan. With the arrival of the Europeans the !San were pushed deeper still into the Kalahari, to areas too dry for farmers and too waterless for herders and ranchers.

The Hadzabe we met in 1958 couldn't imagine a future any different than their ancestors', or care for one. Few had experienced the miracle of turning on a tap to produce clear flowing water, flipping a switch to turn night into day, speaking into a phone to people over the horizon, or listening to a radio broadcasting news and entertainment from around the world. The marvels of modernity the Western world takes for granted were still in their future.

Transposed from the old world of the Hadzabe to the new world of San Diego half a century later, I marvel at Carissa and Guy connecting so seamlessly with their friends back in Kenya by WhatsApp and Facebook. Their globalized world of instant communications comes as naturally to them as hand gesturing among the Hadzabe out on a hunt. For my part, growing up as I did without a telephone or TV, and for a long time no running water or electricity, I am awed by the instant connections between everyone and everything created by the World Wide Web. Zooming in on Google Earth, I can check the glacial melt on Kilimanjaro, look up restaurants in New York on Street Finder, and hold weekly Skype meetings with my Amboseli colleagues when I'm in California. I no longer spend ten minutes rifling through reprints in dusty box files on my library shelves to check a reference. With a click on Google Scholar, I can haul up a paper on my screen in seconds and find cross-references to dozens of related articles. Back in Nairobi, Shirley and I can stream a movie on Netflix while listening to the hyenas and lions in the background. Just as the few Hadzabe still hunting and gathering can't conceive of the internet, so Generation Z kids in San Diego can't imagine the Hadzabe hunting antelopes for food, digging in the sand for water, rubbing sticks to light a fire, or living in a hut made of saplings and grass.

Looking into why the world went flat, Thomas Friedman sees globalization being driven by no one group and everyone in general. He lists a series of modern drivers tied to specific events and dates that flattened and homogenized the world: November 9, 1989, when the Berlin Wall fell; August 9, 1995, when Netscape went public; November 9, 1999, the launch of the open standard HTTP and HTM, which made emails and internet browsers interchangeable.[2]

Hewing closer to the Bretton Woods view, William Lewis, founding director of the McKinsey Global Institute, considers the driving force of globalization to be the efficiency of production—the output per unit of labor.[3] Lewis discounts education, capital, social objectives, and big government. Instead, he singles out competition between companies as the stimulus to growth and economic prosperity.

Important as competition and efficiency are, they don't fully explain our global spread and impact. We are little changed biologically from the first modern humans, so why has it taken over three hundred thousand years to become as efficient and productive as we are in our digital age? To make a Silicon Valley geek out of a Hadzabe hunter calls for fundamental changes in the organization, knowledge, and skills of a society, and above all in its culture.

In his analysis of tax records over decades in the United States and Europe, French social economist Thomas Picketty points to knowledge and skills as the drivers of economic wealth, and investment and training as catalysts.[4] Knowledge and training have unleashed a wave of creativity that has lain dormant in humankind for millennia, but millions remain trapped in poverty with little or no access to the education, health care, and other social services underpinning human development. Pastoralists in Kenya lag far behind farmers, and farmers behind Nairobi residents in access to schools, health clinics, social services, and jobs. Africa's economic growth stagnated from the 1970s to 1990s at a time of rapid acceleration in Asia. Malaysia, at parity with Kenya in GDP when Shirley and I paid a visit in 1982, now has a GDP seven times greater than Kenya's.

Knowledge and skills may be the engine of economic development, yet they are still a step removed from the elemental forces driving our ecological dominance and the transformations foreshadowing globalization. The leaps in innovation, productivity, and efficiency of the global age are possible only once we are liberated from scratching out a daily living and churning out offspring in proportion to available food and when we are freed enough from the clutches of nepotism and cheating to forge large, collaborative alliances and effective governance structures locally and globally. Only then can we lengthen our investment horizons and scale up from local resource-bound communities to a global market economy.

Amboseli gave me a small window into the transition as I followed Parashino, Kerenkol, John Marinka, and other Maasai families from their tightly knit livestock economy and culture into the national and global economy. Like the Hadzabe hunter-gatherers who guided us on elephant hunts in the 1950s, the Maasai a decade later in Amboseli selectively adopted innovations that bettered their lives and standing in traditional society. Lakato was quick to see the benefits of boreholes, dams, cattle dips, and inoculation campaigns in increasing the foraging range and health of his livestock, and of medical services in aiding the survival of his children. Within two decades of government services filtering into southern Maasailand in the 1940s, livestock numbers had doubled. Like most other Maasai, Lakato found that enlarged herds made him better off in what mattered most to him—milk and meat, wives and children, social standing and prestige.

When I arrived in Amboseli in 1967, livestock numbers were topping out at the new ceiling imposed by water development and pasture availability. Over the next twenty-five years I monitored the oscillating numbers until a steady decline set in around 1990 as the grazing pressure of livestock built up and pasture production fell. The 1970s drought that first drew my attention to the conservation knowhow of the Maasai turned out to be the tipping point in their livestock economy and, ultimately, their culture. Such biting drought would ordinarily have curbed family growth, were it not for the surplus American grain stocks handed out by Kerenkol severing the ecological links between rainfall, grass production, and herd size—the age-old erematere link. Once these were uncoupled, the Maasai population rose steadily above the livestock carrying capacity, doubling in twenty years and doubling again in another twenty. Livestock holdings fell to a quarter of the numbers needed to sustain a family on milk and meat. By the 1990s few families could subsist on their herds any longer and, like the family ranchers of Malpai in the American Southwest, had to find other jobs to make do.

Herders who saw the future and embraced change, Simon Salash and Kerenkol among them, gained a front-runner advantage in securing the best lands for farming, starting businesses, educating their kids at private schools, and placing them in good jobs among the new urban professionals. For the poorer families, a way of life and the social bonds binding their society were breaking up. Caught in the transitional vortex between traditional and modern societies, pastoral communities were ill prepared for the rapid changes.

The lure of new opportunities was summed up well by Jonathan Lebo when Shirley and I hosted a large gathering of Leakey Foundation trustees out from California to celebrate Mary Leakey's fifty years of prehistorical research in Africa. The trustees pressed me hard on the impact of my conservation work. Don't you feel bad, they asked, that the wildlife tourism you've promoted is destroying the Maasai way of life?

Rather than give my own view, I invited John Marinka and Jonathan Lebo, both well educated and at the crossroads of the two worlds, to meet the trustees around the campfire in the evening. The Leakey trustees began by saying how much they admired the Maasai and their noble way of life, and how they hoped the Maasai would resist changing to a Western lifestyle. Jonathan Lebo listened politely and answered: our way of life may seem noble to you, but isn't it because you want us to stay unchanged so that you can admire our colorful dress and photograph us? What you don't see is how hard our life is. Every day our routine is the same. At the end of each day we are tired and hungry. We worry about droughts, disease, wild animals, and feeding and raising our kids.

Making little headway, Jonathan tried another tack. What do you do, and you, and you? he asked around the table. One answered forester, another Napa Valley vintner, another Los Angeles property developer, another chemistry professor, and yet another lawyer. Jonathan smiled. Each one of you is doing something different, he pointed out. You can be whatever you want to be, do whatever you want to do—including coming here on safari to see wildlife. Those are the choices we want. Most of us will choose a new life, he added, but some will remain as they are.

Jonathan's parents' goal in life was to have large herds and many children. Unwittingly, like Lakato, they were on the treadmill of change when they used vaccines to boost their herds and medicines to increase their children's health and survival. The *Lebo Effect,* as I dubbed it, accelerated with the schooling of one or two of their children because of growing government pressure and their own exposure to the outside world. Emancipated from ecology and cultural ties, Lebo and Marinka were members of the first generation to choose new lives. Kerenkol's son Josh, off in America, no longer frets about his herd surviving droughts or struggling with the thirteen kids his father is raising. His education and worldly experience offer him a choice of careers and ways of life. Josh has become a different

sort of nomad, migrating to America to get the best education and training before returning to Kenya to land a prime job, marry, raise a small
family, and enjoy the comforts and security of modern life. The options
opened by the early entrants into the growing Kenya economy widen the
gap between rich and poor. Monetary wealth breaks the link between
rainfall, herd productivity, and social relationships. This new wealth, no
longer measured by the size of a herd, weakens the social networks spanning clans and reaching across sections. Herders who remain dependent
on their animals with no other source of income are marginalized in the
new economy. Women are doubly marginalized in having less access to
jobs and having to stay home to raise children.

Lebo's and Marinka's head-start advantage notwithstanding, the early
industrializing nations of Europe and America make it hard for later entrants in Asia and Africa to catch up. The front-runner advantage creates
path dependency, as economists call it. Knowledge, skills, and economic
and political power give older, established industrial nations a huge advantage over later entrants. Colonial American grievances leading to a revolt
against imperial Britain over the trade regulations and excise taxes in the
eighteenth century mirror the grievances of the poorer nations over the
asymmetries of globalization today.[5] Having become a superpower itself,
America is using its economic and military dominance to promote a global
order and trade rules that are to its own advantage.

In contrast to Friedman, globalization scholar Manfred Steger argues
that neoliberal and corporate interests, bolstered by the Bretton Woods
successor institutions, have promoted financial markets that favor the
United States.[6] Steger's claim about the trade and social inequalities caused
by globalization is correct, if not his characterization of the founding
principles of the Bretton Woods institutions. The economic theory that
free trade lifts all nations overlooks historical accident, natural resource
endowment, government policy, and the self-interest of wealthier nations.
Rich nations have selectively lowered barriers and protected jobs at home,
creating wealth gaps between nations and poverty traps within them.

Free trade, blind to front-runner advantage and historical disparities,
typically widens the inequalities caused by globalization and deepens the
discontents.[7] Robert Gilpin points out that domestic and international
markets don't just spontaneously appear.[8] The conditions for an open,

stable international economy must be set and agreed upon impartially, the terms fully implemented and enforced, and the faulty assumptions of parity in knowledge, skills, and investment redressed.[9] The future lies more in soft-power nudge and collaboration over trade and international governance than in military muscle.[10] Howard Rheingold, in *Smart Mobs,* looks at the forces driving the bewildering speed of change in communications.[11] For Rheingold, communications networks grow exponentially in proportion to the number of users. Once connected globally, the networks transcend national and ethnic boundaries. Through international treaties, agreements have been struck to regulate broadband wavelengths, aviation routes, shipping lanes, and other public utilities to avert a global tragedy of the commons in communications and transportation. The rules for managing the commons are similar across all scales from biocultures to global societies, as Elinor Ostrom astutely noted.[12]

Unlike biocultures, in which customary rights and responsibilities are tightly linked to the sustainable use of the land, the globally networked society has become a Faustian bargain—great visible deals on the surface, but with underlying remote and intangible future costs. University of California, Berkeley economist Robert Reich puts it succinctly: "We might make different choices if we understood and faced the social consequences of our purchases or investments *and* if we knew all other consumers and investors would join us in forbearing from certain great deals whose social consequences were abhorrent to us. But we are unlikely to make the sacrifice if we think we'll be the only consumer or investor who refrains. Lonely forbearance can be the last refuge of a virtuous fool. The only way for the citizen in us to trump the consumer and investor in us is through laws and regulations that make our purchases and investments a social choice as well as a personal one."[13]

The disruption of small communities in the rearguard of globalization was foreseen by the American educator John Dewey, who as early as 1927 noted that recent economic and technological trends implied the emergence of a "new world" no less noteworthy than the opening up of America to European exploration and conquest in 1492.[14] For Dewey, the invention of the steam engine, electrical generator, and telephone were breaking up the relatively static and homogeneous local community life predating the global age.

Dewey suggested that the shift from rural farms and villages to large urban centers posed a challenge to open democratic societies. Shades of the pastoralists' world in Africa today, he saw small-scale political communities such as New England townships becoming disrupted by a diffuse amorphous society always on the move, searching out greater opportunities than ever before. He foresaw the disruptions causing political marginalization and social dislocation. The marginalization has deepened political fissures in the West as the new right-wing movements in America, Poland, Hungry, and across much of Europe tussle with liberal parties on the left over free trade, globalism, immigration, and ceding local power to big governments and international bodies such as the World Bank and the UN.

How can citizens come together and make collective decisions, given contemporary society's mania for motion and speed, which make it hard for people to get acquainted, let alone agree on common concerns? Dewey asked. Nearly a century of his concerns are echoed in the growing anxiety on whether America, having accelerated the global express to unnerving speeds, is itself too big to govern.[15] Harvard professor Robert Putnam, in *Bowling Alone: The Collapse and Revival of American Community,* is somewhat optimistic that American communities will revive, if more along the lines of common interests than neighborhood.[16] Authors James and Deborah Fallows, on a long journey into the heart of America, take heart in the towns across the nation forging ahead on urban renewal and innovative developments despite the national dysfunction.[17]

Dewey had no premonition of the IT age and the globally networked society; he couldn't have known the impact they would have on autocracies and democracies alike, or how globalization and the World Wide Web would break down barriers and reconnect individuals around the world and across all domains from commerce to politics, sports, science, and conservation.

I am at once a participant in and observer of one of the last vestiges of a subsistence society boarding the global express. The Lebo Effect—the lure of modernity—and the choices it offers subsistence societies give me reason for optimism. Despite its leveling force, globalization is both inevitable and our best hope for bettering our lives and sustaining planetary health.[18] All the same, I despair at the loss of cultural and natural diversity as the tendrils of globalization invade the most remote and marginalized societies, like that of the Bushmen of the Kalahari and the Pygmies of the Ituri Forest, creating look-alike cities, societies, and landscapes around the world. I worry, too, about the breakdown of the erematere links between family, society, food production, and the health of the environment.

My worry about the breakdown of small-scale, tight-knit societies closely tied to land and neighbors brings me back to the same questions again and again. How different is the global age from earlier human societies? Is the conservation of our rich cultural heritage and biological diversity up to the challenge of globalization, or are we up against changes far too abrupt and disruptive to understand and manage?

14

Our Novel Age

When I began my work in Amboseli in 1967, I spent weeks bashing through the bush in a beaten-up Land Rover tracking down the migrating herds. Now I pick them up from my plane within half an hour or trace their meandering paths by satellite collar beaming down their location every half hour to my home computer. Soon I will send up drones programmed to plot the herds regularly at a fraction of the cost.

Do I miss the day in, day out routine of monitoring animals and plants? Not really. Days slogging through the bush are backbreaking. Recording every observation with robotic discipline is tedious. Science is a discipline, and to be good at it I must measure and report the goings-on in Amboseli rigorously and consistently month after month for years on end. I owe much to the scientific method and consider it the gold standard of documenting nature, testing ideas, and building a knowledgeable society. My dues paid, the routine robs me of the freedom to indulge my curiosity. I prefer to follow my nose, act on a hunch, and watch animals leisurely for the thrill of it. I feel the pulse of Amboseli more sensitively by walking with animals and sensing the energy of the herds than logging their position and numbers from car and plane. I relish seeing Amboseli through the eyes of a cow or a wildebeest and using my natural senses, emotions, and experience as my guide. I learn more and am happiest exploring and savoring nature and humanity rather than dissecting and measuring them.

I saw no hope of saving wildlife in retreating to a quiet grove of acacias and turning my back on the mounting problems in Amboseli. Had I done so, I would have shrunk into ever-remoter spots and done nothing for the animals or people I cared for. I enjoy reading Henry David Thoreau's lyrical writing about his two years in the Walden woods and his poignant message, "In wilderness is the preservation of the world," but he offers no solutions for saving the nature he loved.

Kerenkol looked ahead and figured out how to advance his family and age mates as his world changed. I did the same when the future looked grim for Amboseli by battling my reclusive nature and stepping up to the challenges rather than opting out. Like Bill Conway in his efforts to turn the Bronx Zoo into a conservation center for educating the public, breeding endangered species, and saving the last of the wilds, I've had to rethink and revise my research and conservation priorities time and again to anticipate changes and find a place for wildlife.

To advance my cause, I've taken on roles I never imagined and others I never wanted. From hunter to animal lover to researcher, planner, national director of Kenya's wildlife, and international conservationist, I've reached far beyond my comfort zone, inclination, and skills to work with ranchers, farmers, businesspeople, scientists, and politicians, among them people who eschew my views. The flattening, truncating, and accelerating forces of globalization are bewildering, unsettling, and foreboding—or breathtaking, exciting, and promising—depending on your profession, interests, values, and disposition. Most of all, your views of globalization depend on whether you are aboard the global express or left standing at the station.

The shifting ecological forces of the Human Age are far harder to discern than the changes in society and economy. This is no surprise. We focus on the quick rewards of enjoyment, spiritual relief, and well-being in nature rather than the slow cumulative damages to our planet. We are still hampered by myths and legends about our role in nature and a reliance on other species as models for our own behavior. We are loath to accept that we've broken the shackles of evolution and ecology in so many ways yet are still tethered biologically by the most basic urges of sex, nepotism, tribalism, hedonism, jealousy, and revenge—and ultimately by planetary limits. The new terrain of the Human Age is alien. We have no template for conserving a world so radically different from any other geological or evolutionary era. Biologists have been slow to accept that humans are the new driving force of evolution, or that humanized landscapes and ecosystems are the new norm. Doing so seems to admit a defeat of nature and a godly role for our species.

Unclear about what we have become and where we are headed, we look on life through the rearview mirror and faddishly use the Pleistocene as a benchmark for the healthiest diet, best sex, right exercise, and our ideal social relationships—as if nothing about us had changed in the last twelve thousand years. In doing so, we ignore the few yet crucial genetic mutations and the malleable cultures that distinguish us from other species. In the information technology age, our plastic minds and behavior are changing faster than ever before, molding the world in ways we never imagined, lying distant from our orbit of experience and control.

If we adhered to our ancestral Pleistocene biology and behavior, we would produce as many children as we could. Our choice of partner would be through an arranged marriage decided by parents and elders; we would hunt and track deer and antelopes rather than buy a porterhouse steak at a supermarket, and eat bitter fruits and vegetables so sparse, scant, and bereft of calories that we would spend all day gathering food. As moderns we can fill our stomachs in half an hour with food ten times more calorific and far tastier, thanks to domestication and artificial selection.

Marlene Zuk, in *Paleofantasy*, does a fine job of debunking our misplaced beliefs in bygone Stone Age hunter-gatherers as templates for a fitter, healthier, and happier life.[1] Another modern myth in need of debunking is that we are an ill-adapted misfit of nature, even though our anatomy, physiology, mind, and culture made us superior runners, hunters, and gatherers outcompeting and outsmarting all other species in our African birthplace and eventually around the world.

The time has come for us to confront the seismic shifts created by our ecological emancipation and superdominance and consider how these changes are reshaping ecology and evolution. We have enormous new options to better our lives and confront the awesome challenge of sustaining the health of our planet. We must explain these seismic shifts in behavior if we are to grasp the similarities and differences in the Human Age we have created from our survival of past ages. Economists, biologists, and evolutionary psychologists must look at human behavior as it plays out in daily life, rather than repeat the naturalistic fallacy of philosophers seeking what ought to be from what appears to be natural in nature.[2] Understanding the shifts and implications is a Herculean task, but one we can start on by asking how the Human Age compares with earlier ages. Are its challenges the same as ever, writ large, or do they differ in novel ways?

Political scientist Manfred Steger notes that globalization studies cut across disciplines and are in an embryonic stage. The task is to synthesize the strands of knowledge in a way that allows us to grasp the big picture in a fast-changing world. "Such a trans-disciplinary enterprise may well lead to the rehabilitation of the academic generalist whose prestige, for too long, has been overshadowed by the specialist."[3]

We must shift gear from rearguard actions and mourning bygone eras if we are to anticipate and plan for developments in the Human Age and shuck the muddled views of humankind as intrinsically destructive on the one hand and possessing an inborn love of nature on the other. Instead, we should look on conservation as both the root of our survival and the foundation of our future.

Humanity's growth and spread have been propelled by the same evolutionary forces that drive and constrain all life. Our success can be gauged by our ever-expanding food web, our domestication of resources and environments, and ultimately by the manufacture of material artifacts that make our lives safer, more comfortable, and richer than our ancestors ever conceived. A recurring pattern tracks every step. New stresses and opportunities show up with population increase and economic growth, and old social bonds break down, replaced by more extensive and inclusive bonds. Tribal institutions give way to far larger nation-states, and national governments reach out to forge global agreements and institutions governing trade and investment, setting universal goals, mediating disputes, and addressing global threats like overfishing, pandemic diseases, and climate change. The same human frailties dog every transition and expansion from local to global: selfishness, nepotism, tribalism, hedonism, oppression, and inequality. The frailties emerge strongest during the rootless inchoate transition from customary to modern civic law and political institutions. No sooner do we solve one challenge than our breakneck pace of changes creates new unimaginable and ungoverned ones.

Despite the resurgence of nationalism in many parts of the world, we are on the threshold of a global society, struggling with our dual nature of cooperativeness and selfishness. Joshua Greene, in *Moral Tribes,* points out that it is hard to avoid our tribal perspectives, so we must shift to an inclusive way of thinking about humanity and new common values.[4] Biologists are often loath to venture into the complexities of globalization, which they feel are better left to economists, sociologists, and political scientists. Lost for an explanation like many biologists, I too tend to fall back on other species as analogues for humans, and the past as prologue to the future. There is a reassuring continuity and consistency in the way nature works, and a temptation to apply the same rules to humans as if we were an invariant species.

This standard view, that the same elemental forces are at work in all species, makes sense of the ecology and husbandry strategies employed by members of small-scale subsistence communities like the Pygmies, !Kung, and Hadzabe, who build and inherit few capital assets yet enjoy an affluence in their own way that money can't buy and materialism can't replace. Their core values are the shared social values: the social skills and cultural glue that enabled us to scale up and expand our ecological niche from the world of the hunter-gatherer to the supermarket and eBay forager.[5]

Cultural change is happening so fast relative to evolutionary time that we are making do with the same genetic blueprint as our first intercontinental Paleolithic ancestors, with small but all-important modifications. The big differences between our distant ancestors and us are the technological innovations and changes modifying every planetary process governing and regulating life and the biosphere. Dutch geologist, atmospheric scientist, and Nobel laureate Paul Crutzen, who warned early on of the dangers of a nuclear winter from atomic warfare, ozone thinning, and climate change, first proposed the now widely used term *Anthropocene* to define our human-made age.[6] Crutzen's criteria for use of the term covered humans using one-third to one-half of the land surface; damming or diverting most of the world's rivers; manufacturing more nitrogen than is produced naturally; removing one-third of the primary production of the oceans; exhausting half of the world's readily accessible fresh water; and altering the composition of the atmosphere.

Crutzen's proposal provoked a heated debate about whether our footprint is distinctive enough in the sedimentary record to merit designation of a new geological era. When all we have as a reference point is the sedimentary record and fossils, which accumulate over millions of years, geologists have no choice but to use distinctive markers to define geological eras. A good example is the use of the iridium layer laid down by the Chicxulub asteroid that exploded into the ocean off the Yucatan Peninsula in Mexico and heralded the Age of Mammals. An Anthropocene analogue is the layer of radioactive fallout from the nuclear tests of the 1940s and 1950s, which geologists propose be the benchmark for the start of the Anthropocene. Surely, though, future historians, archeologists, and biologists will use a broad spectrum of indicators to track the rise of human superdominance, not the detritus of nuclear blasts.

Given the welter of changes since the dawn of humankind, I consider geological criteria too coarse to measure our varied impacts around the globe using sedimentary evidence. The scientific debate speaks once again to our boxed-in disciplinary views of the world. Our rise to ecosystem dominance, the extermination of megamammals, and the transition from the Stone Age to the Iron Age and the industrial age happened at different times around the world, making it hard to define distinctive markers of a human-dominated age.[7] Our rich archeological and historical records since the Ice Age enable us to track our scattered imprint from the predawn of our species to the new Human Age.

The imprint we've made on species and habitats over the last few hundred years has been so large as to be dubbed the beginnings of a sixth extinction, right up there with the five earlier mass extinctions, the first in the Permian era 299 million years ago and the fifth, at the end of the Cretaceous, resulting in the demise of the dinosaurs 66 million years ago.[8] As biologist Stuart Pimm of Duke University and colleagues have noted, current rates of extinction are one thousand times the geological average background rates.[9]

As important as they are, human-caused extinctions are dwarfed by the impact of our domestic plants and animals on ecosystems. Dubbed the *New Pangaea* (Pangaea being the Earth's single great landmass before its separation into continents), the intercontinental exchange of corn, rice, wheat, potatoes, bananas, oranges, dogs, cats, cattle, sheep, goats, pigs, rabbits, mice, and hundreds of other plants and animals has created look-alike humanscapes around the world.[10] Our impact has devastated natural habitats, eroded soils, polluted waters, and displaced or exterminated myriads of species. The biological impoverishment applies to our domestic plants and animals no less than wild species. As Russian agronomist Vavilov recognized, we select for specific traits in our domestic species, and in the process inadvertently narrow the genetic diversity of crops and livestock.

The invasion of exotic species choking out native vegetation adds to the biological impoverishment. The *Encyclopedia of Biological Invasions* notes that far more nonindigenous mammals, birds, amphibians, turtles, lizards, and snakes are brought into the United States each year than the country has native species in these taxa.[11] Nonindigenous invaders such as the Asian carp have flourished in American rivers and threaten the survival of native species. A new species invades Hawaii every month, compared to one every thirty-five thousand years prior to human colonization, and in California one every two months, compared to one every ten thousand years before human settlement.

We have directly transformed 43 million of the 80 million square kilometers of ice-free land for crops and pasture, cities, reservoirs, timber, mining, and quarrying. Of the remaining 37 million square kilometers, two-thirds are natural though not necessarily virgin forests, the rest high mountains or tundra. Biomes, the major vegetation zones of the world (such as tropical rain forest, deciduous woodlands, and prairies), have been transformed to human-modified habitats, or *anthromes,* eighteen of them in all, according to Erle Ellis and Navin Ramankutty.[12] The anthromes range from urban areas (eight hundred thousand square kilometers) and irrigated farmlands (1.6 million) to populated forests (7.1), totaling 62 million square kilometers in all. The remaining sparsely used lands include the Amazon, tundra, and great deserts—the wildlands. Even these outwardly virgin lands have been extensively altered, particularly the world's grasslands, deserts, and coral reefs.

Our impact on the evolution process itself has been scantily documented but will be the most enduring imprint of all. We are creating a new evolutionary force through artificial selection, as Darwin called our controlled breeding of dogs, cats, pigeons, and other pets and farmyard animals to produce the traits we favor. We are selecting genes and molding the behavior of domestic plants to produce bigger and richer cereals, fruits, and fibers, and animals to produce more meat, milk, and wool. Our indirect selective forces acting on wild species are fast overhauling natural selection. Whether it is our attack on foxes, coyotes, crows, and baboons when they turn into urban pests; the extermination of keystone species like the wolf and elephant, which created a cascade of ecological changes; or our disruption of nutrient and hydrological cycles, we are altering every ecosystem, habitat, and species. Our impact reaches from the thousands of starlings flocking into the farmlands of Europe and destroying crops to polar bears in the Arctic switching from hunting seals to a diet of geese eggs as the pack ice retreats with global warming.

Conserving species other than for food, as pets, for display and entertainment, or as cultural icons was rare in earlier ages. Natural habitats were seldom protected other than as hunting grounds for royalty and spiritual sanctuaries. Returning to Charles Lyell's view of the economy of nature driving all species, what explains our modern sensibilities of the importance of conserving biodiversity, regardless of a species' utility or threat? Given that our genes have barely changed since the Pleistocene, what can explain the momentous shift from our exterminating species, biodiversity, and natural habitats to conserving them?

Darwin held a pluralistic view of evolution. He recognized that we share most features in common with other animals yet struggled to explain the differences, especially in our moral sensibilities and altruism. Civilizations arose and advanced from the pitiable and rude state of the Tierra del Fuegians he observed on the voyage of the HMS *Beagle* to what he deemed the pinnacle of humanity—his own Victorian British society. As noted earlier, Darwin saw the roots of morality and altruism in warfare, in the courage and sacrifice that created tightly knit cohesive societies, much as Ibn Khaldun did in 1377 and Peter Turchin would amplify in *War and Peace and War: The Rise and Fall of Empires* a century and a half later.[13]

Reflecting on the conditions that favor progress, Darwin wrote: "As man gradually advanced in intellectual power, and was enabled to trace the more remote consequences of his actions; as he acquired sufficient knowledge to reject baneful customs and superstitions; as he regarded more and more, not only the welfare, but the happiness of his fellow-men; as from habit, following on beneficial experience, instruction and example, his sympathies became more tender and widely diffused, extending to men of all races, to the imbecile, maimed and other useless members of society, and finally to the lower animals—so would the standard of his morality rise higher and higher. And it is admitted by moralists of the derivative school and some intuitionists that the standard of morality has risen since an early period in the history of man." Darwin ends with a wonderful comment: "I would as soon be descended from that heroic little monkey who braved his life to save his keeper from the attack of other monkeys or the baboon that rescued an infant from dogs than the savage who delights in torturing his enemies, makes blood sacrifices, practices infanticide without remorse, treats his wives like slaves, knows no decency and is haunted by the grossest superstitions."[14] In other words, Darwin viewed morality as the hallmark of humanity and credits its origins to our animal ancestors, rather than invoking divine origin or condoning superstitions and savagery.

Frans de Waal, like Darwin, sees in human altruism and cooperation a ray of hope for our global age: "I derive great optimism from empathy's evolutionary antiquity. It makes it a robust trait that will develop in virtually every human being so that society can count on it and try to foster and grow it. It is a human universal."[15] De Waal further notes that empathy builds with proximity, similarity, and familiarity, which fits with Darwin's view that empathy evolved to promote in-group cooperation. De Waal states that the greatest happiness is reported not in the wealthiest nations, but in those with the highest levels of trust. In ending, he says that Adam Smith has been misconstrued in invoking the invisible hand as the sole market force. Smith saw virtue, honesty, morality, sympathy, and justice as essential ingredients of a market economy stemming from strong communities.

These views reinforce my point that altruism expands with greater freedom from want, the rise of modern sensibilities, and our recognition of the rights of other people. It extends to a consideration of other species we relate to in one way or another, and ultimately to all species regardless of their value to us. Political philosopher Michael Sandel remarks that "altruism doesn't diminish with use, it grows." He quotes Aristotle, who saw virtue as something we cultivate with practice: "We become just by doing just acts, temperate by doing temperate acts, brave by doing brave acts."[16] Rousseau held the same views in the *Age of Enlightenment:* "As soon as public service ceases to be the chief service of the citizens, and they would rather serve with their money than with their persons, the state is not far from its fall."[17]

There are three other links in the chain from ecological emancipation to our concern for the well-being of other species and the fate of our planet:

the rewards of curiosity, creativity, and doing unto others as we would have them do unto us—the Golden Rule maxim of all major religions and human reciprocity in general.[18] Curiosity and creativity drive us to learn more about the world around us and relate our discoveries to others—we are storytelling animals.[19] Our reward is the emotional buzz of exploration and discovery. Scientists, artists, writers, and poets share the passions of curiosity and creativity. We also get a buzz from doing good works. We are primed to reach beyond our natural evolutionary mold and find reward in the recognition and acclamation of our peers. The works of art, literature, music, and the new inventions and scientific discoveries we create are proxies for our age-old drive for social standing.

Like others, I am a hybrid of my evolutionary past and the global age when it comes to my outlook on nature. I feel a deep satisfaction in the wilds—a product of my genes, biologists who believe in biophilia would say. A better explanation is my upbringing and the wealth of views and experiences I've had since. Culture, not genes, explains my love of animals, the feeling that wells up as I watch the restless energy of countless wildebeest migrating across savannas. I can't imagine an explanation based solely on evolutionary theory. I run counter to our ancestral propensity in giving up the hunting urge with my first kill and appreciating a wildebeest beyond its meat and hides value. In my lifetime I've traced the transition from rural to urban mindset that gave rise to the new sensibilities and values we ascribe to nature—something inconceivable to our ancestors who exterminated the Pleistocene megafauna, or to the American settlers who decimated the bison and put paid to the passenger pigeon.[20]

We need to think of the Anthropocene not as a layer of radioactivity detritus but as a new evolutionary age defined by human superdominance. The Human Age calls for synthesizing the new findings of biology, social science, economics, and psychology to paint a portrait of human nature that represents not what we were, but who we have become.[21] The Hadzabe hunting as their forefathers did and the Maasai hoping to emulate the life of Lakato offer no future for their children, much less their grandchildren. The Human Age calls for adding the best of our inherited traits and lessons to the best of contemporary knowledge if we are to preempt the unintended and unanticipated consequences of our superdominance. Surely we are aware and skilled enough to figure out how to expand our senses and analytical powers through technological proxies, and to hone our cultural facilities to domesticate our baser instincts and curb our maladaptive evolutionary urges.

Each innovation and each revolution, from the Late Pleistocene tool revolution to the Neolithic domestication of plants and animals to the growth of city-states and nations, has accelerated the pace of change and our ability to transform the world to our own ends. Our new global society has created a different ecological and evolutionary epoch than any before. Natural diversity is declining even as the variety of human-manufactured items, musical productions, literature, architecture, and other cultural artifacts is climbing. Cities are the most biologically impoverished anthromes yet the richest in human creativity.

Globalization retains the same basic ingredients as all revolutions. Self and family interests along with our ability to cooperate and scale up to ever-larger societies are universal human traits reflecting our evolutionary past. Opportunity, wealth, security, social advancement, and approbation still drive us. These deep-seated forces create the same tensions between cooperators and cheats as in small hunter-gatherer societies, though with far greater disparity in wealth and power. Self-interest, nepotism, tribalism, cronyism, culture, and religion still divide us, if with less intensity in some nations than others. Free riders, despots, and warmongers still undermine progress. The rapidity of change, social mobility, and the weak connections between action and consequence slow social and ecological feedbacks in the global age, making it harder for us to detect and curb our harmful behavior.

If globalization has left a large lacuna in the transition from local ecosystem to planetary limits, are there surrogates for proximity we can use to reconnect the biocultural links and feedbacks? How can we remake ourselves, our culture, and our husbandry skills fast enough to meet the global challenges of the human-dominated age?[22]

15

The Modern Conservation Movement

A visit to Yellowstone had long been on my travel agenda before I finally made a pilgrimage to the birthplace of national parks in 1978. Famous for its cauldron of hot springs and erupting geysers and as the last redoubt of America's bison and grizzlies, Yellowstone was the first land any nation ceded as a national park. Writer of the West Wallace Stegner called national parks the best idea America ever had. Helen of Troy may have launched a thousand ships: since its birth in 1872 Yellowstone has launched seven thousand national parks worldwide.

The first parks in Africa—Kruger in South Africa, Metobo in Zimbabwe, and Akagera in Rwanda—were gazetted in the 1920s and 1930s. Others were added in a quickening pace across the continent in the next two decades, first as bulwarks against the rising tide of farmers, herders, poachers, and developers, and then in a rush in the 1950s, as colonial governments feared that approaching African independence might spell doom for wildlife.

If Yellowstone was the best idea America ever had, how have parks fared in the rest of the world? America was exceptional in many ways. The European colonizers of the New World decimated the Native American population through disease, persecution, and starvation, then forced the survivors into reservations and slaughtered the great migratory herds of bison and pronghorn antelopes to make way for white settlers. The depopulations and exterminations made it possible for the United States to set aside great tracts of land as national parks and forest reserves.

Creating Yellowstone-like parks in Africa was a different proposition that at the outset ignored Africa's own exceptionalism: its stunning variety and marvelous abundance of wildlife living alongside people. Surely another way of conserving wildlife was possible without evicting people from their homelands and without making an Old Faithful spectacle of lions and cheetahs, disrupting their behavior.

After Amboseli was declared a national park in 1974, it shared with Yellowstone the problems of being loved to death. In 1975 I was appointed an honorary planner for Kenya National Parks by the director Perez Olindo and took the opportunity to draw up road plans to disperse the crush of minibuses, spare the harried cheetahs and lions, and curb off-road driving slicing up the grasslands and spewing dust contrails across the park.

My venture in tourism studies was chore rather than charm, but it had its lighter moments. Out measuring the dust emissions from racing minibuses, I laid out a row of paper plates and drove back and forth upwind like a tour driver hunting down lions. Baboons, used to snatching morsels from visitors at the nearby lodge, made a beeline for the plates as I raced after them, yelling like a raving banshee, to retrieve my dusted plates.

In Wes Henry I was lucky to find the right person to take over my visitor studies. A tall and amiable graduate of Colorado State University, Wes was well acquainted with the love-it-to-death problems facing America's parks. I soon had him tracking minibuses and jotting down details of tourists banging on car doors, throwing spent film cartons at sleeping lions, cutting off rival buses, and racing back to the lodges at midday to beat the lunch queues. Like the predators', visitor behavior after dark was off record.

The main visitor attractions in African parks are the Big Five, a term borrowed from the big-game hunters who track down trophy lions, leopards, elephants, rhinos, and buffaloes. Wes's study showed tour drivers spending 80 percent of their time tracking down the Big Five, with cheetahs substituting for leopards, hard to bag on film. He regularly counted forty minibuses milling around the lions, although he found that visitors spent barely enough time to snap a few photos after jostling through the melee. Lions took to using the rows of buses as cover to stalk prey. The day-hunting cheetahs were less fortunate, thwarted again and again by buses gunning ahead for a frontal-attack photo. Several cheetah females with cubs abandoned the park as their kill rates fell with visitor obstruction.

Based on Wes's findings, we came to the startling conclusion that the carrying capacity of Kenya's parks turned on how many visitors could squeeze around a pride of lions, the number of prides in a park, how many visitors a pride could tolerate, and a visitor's tolerance of the congestion. Despite our findings, the Ministry of Tourism and Wildlife kept on licensing new lodges. By the 1980s Kenya's reputation as a wildlife mecca was sullied by the Big Five mania.

Anxious to do something about the tourism tragedy of the commons, I took up a ministry offer to devise a visitor management plan for Amboseli. The government adopted these tourism zoning principles as a model for other parks and, shortly afterward, asked if I would set up a Wildlife Planning Unit to cover parks and wildlife conservation countrywide. I went on to write an article for *BioScience* with Wes Henry laying out a general framework for managing wildlife tourism, reducing visitor impact, and engaging local communities.[1] The framework laid a foundation for ecotourism and the creation of the International Ecotourism Society Megan Epler Wood and I set up in 1991.

Meanwhile, I felt it was high time to visit America's national parks. Bill Conway arranged an official tour for me to meet park superintendents, researchers, and planners in Yellowstone, Grand Tetons, Grand Canyon, Everglades, and other parks across the country.

Thrilled at making it to Yellowstone in the early summer of 1978, I watched awestruck as crowds of sightseers circled the Old Faithful geyser, waiting for it to blow as regularly as a metronome. Further down the road scores of visitors circled another Yellowstone icon, a massive bull moose as tame as a Jersey cow. On the ensuing tours of Grand Tetons, Grand Canyon, Yosemite, and Redwoods, I was struck by the contrast with African parks. America's parks symbolize the grandeur of the country's landscape—the craggy mountains, yawning canyons, and towering forests used as outdoor recreational areas for hikers, campers, anglers, boaters, bikers, climbers, cavers, and skiers. African parks—Serengeti, Amboseli, Tsavo, Ruaha, Murchison, and Queen Elizabeth among them—were set aside as wildlife conservation areas where the visitor is an onlooker confined to a vehicle insulated from nature.

The notion of preserving land to protect wildlife was first touted by George Catlin, artist and early visitor to America's continental interior. In

1832, mesmerized by the vast herds of bison living in seeming harmony with the indigenous tribes of the prairie grasslands, Catlin urged the country to set aside "a nation's park containing man and beast in all the wild and freshness of nature's beauty. Such a park would preserve and be upheld for refined Americans and the world to view in future ages."[2]

The first step toward a nation's park took a different turn than Catlin envisaged. In 1864 Abraham Lincoln signed an act of Congress ceding Yosemite to California as a state park to protect its spectacular landscape for "public use, resort and recreation, inalienable for all time." Yellowstone, set aside as a national park by an act of Congress in 1872, was also slated for public use and recreation, including hunting until it was banned by the passage of the Act to Protect Birds and Animals in 1894. Like all America's early parks, Yellowstone's status was contingent on no better uses such as timber and mining standing in the way. Nature and wildlife conservation barely featured in American parks until the birth of the National Park Service and an explicit passage in the National Parks Organic Act of 1916: "to conserve the scenery and the natural and historic objects and the wildlife therein and to provide for the enjoyment of the same in such manner and such means as will leave them unimpaired for the enjoyment of future generations."

A further two decades passed before ecological values were finally recognized with the creation of Everglades National Park in 1934. William Hornaday, the conservationist who had done so much to save the last of the bison at the New York Zoological Society, was not persuaded this was a good choice. He saw absolutely nothing picturesque or of conservation value in the Everglades: "A swamp is a swamp, . . . a long way from being fit to be elevated to the magnificent array of scenic wonderlands of US parks."[3]

The story of African parks runs a divergent course, closer to George Catlin's notion of a park preserving wildlife and people living in harmony than the monumentalism of Yosemite and Yellowstone. The earliest move came in German East Africa, now Tanzania, when in 1886 the trans-Africa explorer Hermann von Wissman set aside two reserves to protect the scientific and aesthetic value of wildlife. His extraordinary vision in setting aside the reserves to protect wildlife for its intrinsic values came eight years before hunting was banned in Yellowstone and a half century before America recognized the ecological value of parks.

Sir Harry Johnston, special commissioner to Uganda, wrote to the British Colonial Office in 1894 saying, "It would be melancholy to think that such glorious creatures as the eland, the kudu, the sable antelope and zebra were passing into extinction when they might be saved and perpetuated by our making a little effort in the right direction." Johnston's plea was followed by the foreign secretary convening eight colonial powers in London in 1900 to the first ever international wildlife convention, aiming to curb the excesses of the ivory trade and sport hunting to protect Africa's elephants and other wildlife.[4]

Sir John Kirk, Scottish physician, explorer, naturalist, and British consulate to Zanzibar (among his many extraordinary achievements, he persuaded the sultan of Zanzibar to ban slavery), followed von Wissman's lead by setting aside two vast tracts of land in 1898 to protect Kenya's wildlife. The twenty-thousand-square-kilometer Southern Reserve spanned from Amboseli to Mara. The larger Northern Reserve in the arid Samburu region stretched from the Uaso Nyiro River to the Mathews Range. Urging "a little effort in the right direction," the Society for the Preservation of the Fauna of the Empire pushed the British colonial government to establish a slate of wildlife reserves: "We owe the preservation of these interesting and valuable and sometimes disappearing types of animals as a debt to nature and to the world—we are the trustees for posterity of the natural content of the Empire—the reserves ought to exist not for the gratification of the sportsman, but for the preservation of interesting types of animals."[5]

The starkly divergent visions for American and African parks began to converge only after the World Parks Congress of 1982, where the Amboseli case study, along with a handful of others, paved the way for an ecosystem approach and community engagement. A coalition of conservation organizations under the Greater Yellowstone Coalition banded together shortly afterward with a similar vision for Yellowstone. At a meeting hosted by Art Ortenberg and Liz Claiborne that I dubbed "From Yellowstone to Amboseli and Back," Shirley and I joined the Nature Conservancy, the Wilderness Society, and other conservation organizations at the Gallatin Gateway Inn in the spring of 1992. The converging visions from such divergent starting points was striking. Whereas the Amboseli ecosystem plan aimed to prevent carving up the land and disrupting migrations, the mission of the Yellowstone Coalition lay in restoring the ecosystem and the free movement of wildlife beyond the park boundaries. Winning space for bison, elk, and deer to roam across their former migratory range would, according to biologists, alleviate the need for a controversial culling program to protect ecological damage to riparian habitats.[6]

A big obstacle in the path to a Greater Yellowstone Ecosystem was a powerful pro-development coalition backed by the People for the West and the Wise Use Movement. As far as they were concerned, the prosperity of the American West lay in logging, mining, hunting, and ranching, and that's the way it should stay. They blamed the loss of their traditional livelihoods on the "post-cowboy economy"—new service industries, tourism, rich easterners buying land for leisure, and conservationists claiming timber, mining, and ranching lands for wildlife and habitat.[7] Winning over the ranchers was proving a hurdle for the Greater Yellowstone Coalition proposal and still remains an impediment.

I left Yellowstone realizing the stunning success of the national park movement lay in the evolution of the park concept, not in its original mission to protect the grandeur of America's landscape for public recreation. Robert Keiter, reviewing the history of U.S. national parks a century on, captures the changing Yellowstone vision: "American society looks and thinks quite differently today than it did when the national parks system was created. Better informed by science about ecological imperatives, species conservation, restoration requirements, and climate change, we have the opportunity to reassess the purpose of the national parks and continue to meet these emerging challenges. Ecosystem-based expansions, national restoration areas, multiagency landscape-based national monuments, federal wildlife corridor legislation, better-coordinated ecosystem management arrangements, and minority-focused urban park units can help meet tomorrow's conservation demands."[8]

My trip around America and into its past persuaded me that Yellowstone was no flash of inspiration popping into the mind of some visionary conservationist: the greatest idea America ever had ran far earlier and deeper in the American psyche, giving birth to the modern conservation movement in the early twentieth century. I ended my tour with far more exploration yet to do than I had imagined in first visiting Yellowstone. Why, for example, did the modern conservation movement itself emerge in America and have such influence around the world? How relevant is the American notion of the wilds and nature elsewhere in the world and to the challenges of the global age?

In *Wilderness and the American Mind* conservation historian Roderick Nash gives a fine account of how urbanization and expanding sensibilities changed American views of the wilds in the late nineteenth and early twentieth centuries.[9] From the remoteness of the urbanized East, the new bucolic view of the wilds turned the Pilgrim settlers' forbidding images of the forested interior into John Muir's cathedrals of nature and the detestable wolf into a lovable Bambi.

Oddly, despite the rich historical accounts, conservation itself hasn't been studied as an adaptive and evolving response to our changing world and aspirations. Narrowing the meaning of conservation to equate it with saving habitat and biodiversity masks the many threads woven into the modern conservation movement—and its yet remoter origins. Casting conservation as a modern concept rising phoenix-like in the twentieth century gives the illusion of conservation as an invention akin to electricity, cinema, and the airplane, promoted by the West for the good of a developing world abusing its lands and not caring a hoot about wildlife and nature.

Conservation has become a smorgasbord of human endeavors. Like cuisine and art, conservation appeals to a miscellany of tastes with seemingly little in common. Despite the divisions and angst caused by its very diversity, an underlying unity and continuity runs through conservation. The thread originates in necessity, utility, efficiency, sustainability, and environmental security, evolves to encompass sport hunting, tourism, wilderness, culture, art, and artifacts, and culminates in sustaining the diversity of life and the health of our planet. We can understand the richness of conservation and the role it plays in our lives and the future only by tracing back the threads woven into it.

One thread in the modern conservation tapestry is the same wonder and awe that spurred eighteenth- and nineteenth-century reverence for nature in Europe: romanticism.[10] In the United States, romantic artists in the urban East were drawn to the grandeur and seemingly pristine wilds of the vast American hinterland. Most notable was the Hudson River School of landscape artists who came to prominence in New York in the mid-1800s under the tutelage of English immigrant Thomas Cole. Cole had been schooled in the British tradition of portraying the sublime and fearsome in nature, pioneered in Joseph Turner's brooding rural landscapes. Frederic Church and Albert Bierstadt became the most famous Hudson River School exponents of America's scenic grandeur, a grandeur depicted in their dramatic frescoes of Niagara Falls, Yosemite, the Rocky Mountains, and Yellowstone.

The artists of the Hudson River School brought back to the East enthralling images of the wildness and grandeur of the West. Among its ranks was Thomas Moran, born in the same English town of Bolton as Thomas Cole. Moran joined the head of U.S. Geological Services on a trip to Yellowstone in 1871 and painted the resplendent scenery that persuaded Congress to create America's first national park. For all the Hudson River School's influence, the reverence for nature it espoused soon met the hard realities of the American pioneers moving west to claim land. In their wake came lumber companies, miners, meat hunters, and fur traders exploiting the natural bounty of the West.

Writer-naturalists in the American transcendentalist movement, among them Henry David Thoreau and Ralph Waldo Emerson, added another thread in America's conservation tapestry. Emerson grieved over a nation in transition from a wild continent to a sprawling metropolis.[11] Thoreau, celebrated author of *Walden; or, Life in the Woods*, tired of urban life and retreated to the quiet woods of Walden Pond near Concord, Massachusetts, to "transact some private business with the fewest obstacles." Returning to urban life two years later, he wrote that in nature, "the universe will appear less complex, and solitude will not be solitude, nor poverty, nor weakness," overlooking the urban lifeline that sustained his retreat at Walden.[12]

Yet another thread of the modern conservation movement is the spiritualism of John Muir, the Scottish émigré who, like the Hudson River School of artists, exulted in nature, in his case in the towering cedars and pine forests of the Sierra Nevada. He was among America's staunchest defenders of the wilds.

The animal welfare and rights movement of the late nineteenth century, closely tracking abolitionism and the movements for the rights of women and children in Europe, sewed another thread. Rejecting the mechanistic view of Enlightenment philosophers who viewed animals as machines incapable of thought or feelings, the animal welfare movement lobbied for the ethical treatment of farm animals, beasts of burden, and pets.

Perhaps because of the boundless view Americans took of their vast new continent and its seemingly unlimited abundance, it took them time to wake to the destruction of their lands, forests, and wetlands, despite the alarm raised by influential naturalists.[13] Chief among them was German explorer and polymath Alexander von Humboldt. In a famous narrative of his explorations of South America, Humboldt alerted the world to the destruction of the southern continent by Spanish settlers at the turn of the nineteenth century.[14] He became a close friend and correspondent of President Thomas Jefferson, and his warnings prompted George Perkins Marsh, a U.S. congressman who helped found the Smithsonian Institution, to publish *Man and Nature; or, Physical Geography as Modified by Human Action* in 1864.

Marsh was a man of extraordinary talent and energy. Serving as U.S. minister to Turkey for several years, followed by a twenty-one-year stint as minister to Italy, Marsh toured widely and was stunned by the destruction he saw in the Mediterranean: "There are territories larger than all Europe that once sustained populations scarcely inferior to the Christian world that have been withdrawn from human use, where human action has brought desolation almost as complete as the moon, where in the brief space of time in the historical period, they were covered with luxuriant woods, verdant pastures and fertile meadows."[15]

Marsh broke with the bottomless cookie jar view of nature as boundless and endless, warning that deforestation and soil erosion would ruin civilizations of his day as they had those of ancient Greece and Rome. He was no Cassandra, though. Agricultural technology could keep up with the world's rising population, he wrote, so long as there was soil left to plow. His message would find champions a half century later in the Progressive Era of Teddy Roosevelt and Gifford Pinchot.

Often overlooked in the weave of the conservation fabric are the hunters and outdoors enthusiasts who formed conservation-minded bodies such as the Boone and Crockett Club in 1887. The hunters, fearing the extinction of their game animals, pushed for legislation to protect wildlife, curb commercial hunting, and limit the length of hunting season.[16] Other threads were spun in the nineteenth and early twentieth centuries by naturalists like John Audubon, who celebrated the beauty of birds in his dazzlingly illustrated *Birds of America,* printed in 1827. Other threads were woven with the emergence of the science of natural resource management, ecology, and the birth of nongovernment conservation organizations such as the Sierra Club, founded in 1892 with Muir as its first president, and the National Audubon Society in 1905.

The new sensibilities turned millions of Americans into nature lovers, animal devotees, adventurers, recreationists, hikers, and escapees from the crowded industrial cities at the turn of the twentieth century. The continental scale and speed of transformation from a pre-wheel indigenous culture to the age of rail and air travel in just over three centuries, coupled with the stark juxtapositions of wild and domestic landscapes, made America's conservation movements arresting and forceful.

If one figure stands out as encapsulating the most strands of the modern conservation movement, it is the manly figure of a U.S. president, Theodore Roosevelt, a hunter, naturalist, writer, adventurer, explorer, and cattle rancher imbued with an ethos of living on the range. Roosevelt learned his love of the wild camping in the Sierras with John Muir and in his association with Gifford Pinchot, who had been tutored in European forestry and was influential in founding the Yale School of Forestry. On becoming president, Roosevelt charged Pinchot, a key architect and staunch advocate of Roosevelt's Progressive movement, with setting up and overseeing the U.S. Forest Service.

Unlike the transcendentalist Thoreau and the spiritual Muir, Roosevelt saw the industrial revolution as having given America both "marvelous well-being" and the "care and anxiety inseparable from the accumulation of wealth." The duty of government, Roosevelt believed, was to protect the interests of all Americans and husband the nation's wealth of wildlands and natural resources to achieve "the greatest good for the greatest number." He campaigned vigorously against wholesaling America's public lands to the lumber barons and mining magnates, despite their enormous wealth and power. To protect America's natural heritage for all time, Roosevelt took the extraordinary steps of protecting 93 million hectares of federal land in 150 national forests. He went on to create five national parks, eighteen national monuments, four national game preserves, and the first wildlife refuge to protect an endangered species, the brown pelican in Florida.[17]

The Roosevelt years wove the tangled threads of nineteenth-century and early twentieth-century conservation into a broad tapestry of principles and practices for managing natural resources, wildlife, monuments, and wilderness at the core of the modern conservation movement. The federal and state governments set up oversight agencies run by professional managers and planners guided by policies, legislation, and regulations.

For all the new sensibilities, philosophies, and government programs, the roots of the modern conservation movement grew as much from the pushback against the degradation of land and depletion of resources caused by European settlers. As Adam Smith noted, the colonization of America was a matter of European technology and skills transplanted to a new continent.[18] The pioneers imported their own farming practices based on wheat, barley, oats, temperate fruits, cattle, sheep, goats, chickens, ducks, geese, and horses for transportation and plowing. Adding maize, squash, pumpkin, and other indigenous foods to their familiar repertoire, the colonizers transformed the land more extensively and intensively in three hundred years than the Native Americans, who lacked the wheel, plow, cow, and horse, had in fifteen thousand. By the late 1800s American settlers had razed forests and woodlands, drained wetlands, dammed rivers, hunted bison to near extinction, and transformed the continent into an agrarian landscape.

To the pioneers, America was so vast and virgin as to give the illusion of a wild nature bountiful and inexhaustible. Land stolen from the Indians was appropriated by European settlers, as were the forest trees powering machines, building homes, and heating hearths, and the bison, deer, and antelope herds providing meat and hides. The frontier spawned a culture with no ancestral roots in the land, and no sense of its finiteness and fragility.

Historian Fredrick Jackson Turner, writing in the 1890s, claimed that American democracy was made by the westward expansion of the pioneers opening new lands to settlement.[19] The closing of the frontier brought an end to the American spirit of moving on when resources ran out and the land was exhausted. Roosevelt captured the new reality and growing concerns at the Conference of Governors on the Conservation of Natural Resources in 1908: "What will happen when our forests are gone, when the coal, the iron, the oil and the gas are exhausted, when the soils shall have been still further impoverished? . . . We began with soils of unexampled fertility, and we have so impoverished them by injudicious use and failing to check erosion that their crop-producing power is diminishing instead of increasing." No catastrophe brought attention to the destruction of the land or did more to galvanize conservation action than the Dust Bowl.[20]

The prairies, stretching like an inland sea of grass from the Canadian border to Oklahoma and the Texas Panhandle, supported millions of migratory buffalo when the first white colonizers arrived hungry for land. By then the Comanche, Sioux, Cheyenne, Cree, Kiowa, and other tribes that traditionally hunted bison had acquired horses from the Spaniards and become proficient in keeping up with the herds. Before long, their growing impact on the bison was overtaken by the commercial meat and hide hunters in the largest wildlife massacre of all time. By 1883 the 30 million or so bison on the prairies had vanished, killed for their meat, tongues, and hides, which were shipped east on the new railroads traversing the continent. All that remained were piles of bones that would supply a thriving market for fertilizer, bone meal, and glue for another decade.[21]

With the buffalo gone and the Indians subdued, the prairies lay open to white settlers and cattle ranching. The XIT Ranch in the heart of the Llano Estacado prairie in the Texas Panhandle was the largest spread. Feeding a booming cattle market, the herds grew and the prairies were carved up, fenced off, and overgrazed.[22] Seeing the same signs of degraded grasslands that launched the Malpai Borderlands Group in the 1990s, the XIT ranchers fretted over the frailty of the land. Unlike in Malpai, the rains were good enough to farm the prairies, and so they did. A run of good years, high wheat prices, and cheap land opened by the Homestead Act of 1862 and promoted by a federal government and states anxious to turn the plains to profit, and the promise of a quick fortune lured homesteaders from as far off as Germany and Russia.[23] Over 1.2 million hectares of land were cleared in a few years by sodbusters, as they were called. Cattle herders warned the "nesters," the settler farmers, of the dangers of breaking up the sod: "Miles to water, miles to wood, and only six inches to hell."[24]

The land lay barren during the winter when high winds blew out of the north. The wheat boom withered as wet years ran dry and wheat prices plunged. Farmers plowed more land to make up faltering profits and secured loans to buy farm machinery. By 1934 the denuded land began to lift, creating huge dusters, black blizzards, and drifting sands. The culmination came on Black Sunday, April 14, 1935. A black duster hundreds of kilometers wide and nearly a thousand meters high rolled in from Kansas across northern Oklahoma and Texas like a giant tsunami. The roiling granular dust blinded farmers caught in the open, clogged the lungs of people and animals, and smothered homes and barns in drifting sands.[25]

Displacing over 2 million people, the Dust Bowl disaster helped bring down the Herbert Hoover government, just as the Kongwa groundnut scheme in Tanganyika would topple the Atlee government in Britain two decades later. The new president, Franklin D. Roosevelt, proved as bold as his distant relative, Teddy Roosevelt, when it came to conservation. FDR hired veteran soil scientist Hugh Bennett, who had studied how civilizations abroad had sustained their soils over the centuries, to report on the causes of the Dust Bowl. Bennett had a hard time convincing a skeptical Congress that man, not nature, was destroying the prairies and devastating farms and farmers. Four days after the clouds rolled across the prairies that Black Sunday, the tail end of the dust storm darkened the sky of Washington, DC, as Bennett was addressing Congress. He pointed out the window: There goes Oklahoma. Bennett won support for a restoration plan to buy up millions of head of surplus cattle and sheep, subsidize wheat sales, and purchase failed homesteads. Under the New Deal Roosevelt set up the Civilian Conservation Corps to plant millions of trees and build dams to curb the winds and stabilize the soil.

The Dust Bowl launched a slew of soil and water conservation programs overseen by government agencies and run by professionals, adding to the efforts of the land grant colleges, agricultural extension services, the Forestry Service, the U.S. National Park Service, and other arms of government. The interlocking efforts launched natural resource management as a specialized branch of conservation.

It took one disaster, the Dust Bowl, to galvanize America to conserve its soils, and another, the near extermination of the bison, to shock the nation into saving its wildlife. Early in the twentieth century the virtuous farmer slaughtered predators, aided by state agencies. Wolves, bears, and coyotes were cast as "brutal murderers" killing "harmless, beautiful animals"— namely, livestock. In 1906 the Bureau of Biological Surveys launched a campaign to eradicate wolves and coyotes with traps, poisoned bait, and bounties for every tail. A bureau biologist declared, "Large predatory mammals, destructive of livestock and game, no longer have a place in our advancing civilization."[26]

The threads woven into the Progressive Era and the modern conservation movement converged in Aldo Leopold. Leopold was raised and trained at the height of forest and range destruction and the Dust Bowl era. At an early age he was appointed to government office in the formative years of the U.S. National Park Service, Forest Service, Bureau of Land Management, and Fish and Wildlife Service. An avid deer and duck hunter, resource manager, naturalist, researcher, teacher, writer, and philosopher, Leopold laid the foundation for a new land ethic in America.

Leopold saw humanity not as standing apart from nature, but as its user and custodian, its very epicenter. Out hunting wolves to create more deer for hunters like himself, he had an epiphany watching the "fierce green fire" die in the eyes of an animal he shot to control predator numbers. That fire awakened in Leopold a sense of how nature fitted together as a whole, and the role of the wolf in the ecological balance: "Damage to plant life usually follows artificialized management of animals—for example, damage to forests by deer. . . . In each case over-abundant deer, when deprived of their natural enemies, have made it impossible for deer food plants to survive or reproduce."[27]

Leopold deftly wove together the ancient threads of land husbandry and resource management with the new conservation creed of wilderness and wildlife conservation. Blending aesthetic and ethical caring for the land with utility, his classic book, *A Sand County Almanac*, included the most-often-quoted passage in all of modern American conservation: "Examine each question in terms of what is ethically and aesthetically right, as well as what is economically expedient. A thing is right when it tends to preserve the integrity, stability, and beauty of the biotic community. It is wrong when it tends otherwise."

Drawing on his scientific training, Leopold conceived of his land ethic as dependence on and husbanding of the natural assemblages of animals and plants making up a functioning community. We should keep all the ecological elements intact to maintain the land's capacity for renewal. "A land ethic, then, reflects the existence of an ecological conscience, and this in turn reflects a conviction of individual responsibility for the health of the land."

The land ethic has often been read as a set of laws and prescriptions taught in college alongside farming and forestry techniques. Leopold held a far richer view: "I have purposely presented the land ethic as a product of social evolution because nothing so important as an ethic is ever 'written.' The extraction of value is never automatic; only healthy cultures can feed and grow. . . . There is as yet no ethic dealing with man's relation to the land and to the animals and plants which grow upon it. The extension of ethics . . . is an evolutionary possibility and ecological necessity. . . . Ethics are possibly a kind of community instinct-in-the-making."

I suspect Leopold came to his inclusion of the social dimensions of his land ethic late in life, after laying the foundation of his land-human linkage based on the ecosystem concept of British ecologist Arthur Tansley. Had his life not been cut short, had he read of Vavilov's biocultures, the Konso's fuldo system in Ethiopia, and the erematere concept of the Maasai, I feel sure Leopold would have extended his land ethic to culture.

Although Leopold saw social approval and disapproval as the driving force of ethics, he had less feel for culture in the evolution of ideologies, norms, and customs as its underpinnings. This is not surprising. By virtue of their colonization, the ethos of individualism and conquering nature among American farmers and ranchers is far stronger than in ancient settled Europe, Asia, and Africa. Most Americans lack the cooperative nature and community glue to develop a fuldo or erematere culture, or the community-centric view of Chinese rice farmers.[28] Rekindling the ethos of cooperation and a land ethic takes Leopold's vision, Stegner's call for a society to match its scenery, and the Malpai ranchers' dedication to restoring their land and preserving their cowboy culture.

The worry of environmentalists became the worry of all as pollution chocked the skies and fouled rivers and lakes around the world. In America the growing environmental movement assailed big corporations and lobbied government, ushering in a slate of regulations on emission standards, monitoring, and enforcement under the Air Pollution Control Act of 1955. The Environmental Protection Agency, established in 1970, played a large role in reducing pollution emissions, aided by the polluter-pays principle and class action suits against big business.

The scourge of burning lungs and respiratory congestion in the industrial West eked into an invisible greenhouse mantle enveloping the Earth as huge tankers spewed giant oil slicks, killing marine life and fouling beaches from Alaska to Kenya. Environmentalists today warn of the threats to the Earth's biosphere just as Rachel Carson did of the dangers to America's lakes, rivers, soils, and air a half century ago.[29] Barry Commoner's 1971 best-selling book, *The Closing Circle*, showed that no place in our shrinking planet was safe from our industrial effluents.[30] Over the next two decades environmental activism and the media bullhorn ushered in a spate of international accords: the United Nations Convention on the Law of the Sea, the Commission on Environment and Development, the Montreal Protocol on Substances That Deplete the Ozone Layer, the Convention on Biological Diversity, and the Framework Convention on Climate Change among them. Academics rose to the complexities of the global challenges with new interdisciplinary studies such as environmental economics, conservation biology, and Earth systems science.

The multihued threads woven into the birth and evolution of the modern conservation movement aside, is a movement born in the West in the nineteenth century up to the challenges of human superdominance and globalization in the twenty-first?

On the plus side, the modern conservation movement has evolved and adapted to the realities of our superdominance. As Keiter notes a century and a half after Yellowstone launched the greatest idea America ever had, society looks and thinks quite differently today: we have an opportunity to reassess and meet the challenges of the Human Age.[31]

On the downside, modern conservation has become divorced from its roots in the survival, productivity, efficiency, and resilience of our ancestors struggling for species dominance and ecological emancipation. Leopold put it well: we can be ethical only in relation to something we can see, feel, understand, love, or otherwise have faith in. The extension of ethics from the social realm to planetary health is, as he recognized, an evolutionary possibility and ecological necessity in our global age. To expand from the passion of a privileged few to citizens in all walks of life, conservation must become as deeply ingrained in our cultural values and everyday habits as are personal hygiene and public health. How can that be done using surrogate senses at a global scale to mimic the natural connections of close-knit communities?

PART

III

Our Once and Future Planet

16

Unnatural Reconnections

We are most of the way through a momentous demographic transition, from a peak growth rate of 2.1 percent in the 1960s to a population expected to level out at 10.9 billion around 2100, according to the UN's "World Population Prospects," issued in 2019.[1] Europe and North America already produce mountains of surplus food, and Asia will soon meet its own needs. Africa's demographic transition is finally under way after decades of economic stagnation. Since 1960 annual population growth rates have dropped in most countries—in Kenya from 3.1 percent to 2.5 and in Botswana from 2.3 to 1.8, though the tide is still rising in several poor countries, for example in Mali from 1.1 to 3.0 and in Somalia from 2.1 to 2.9.

Development goals have tilted in most nations from the runaway population concerns of the mid-twentieth century to human welfare and environmental quality in the early twenty-first century. The environmental concerns include air and water pollution, soil erosion, excessive fertilizer and insecticide use, the loss of natural habitats and biodiversity, and, topping the global agenda, the disruption of planetary processes.

The atmosphere, oceans, and terrestrial system interact through carbon, nitrogen, hydrological, and other elemental cycles that, through complex feedback processes, maintained Earth within a narrow bandwidth that spawned a growing diversity of life until the Human Age. Today our global impact is uncoupling the feedback systems and driving a cascade of changes ever harder to track and counter. The dangers of a hot Earth to political and economic stability in the twenty-first century will far exceed those of population growth and food sufficiency in the twentieth century.[2] Views on how to handle the hot Earth threat to the future of our planet range from a comprehensive global agenda to shaking up a lethargic public and apathetic governments with cataclysmic projections.[3]

No nation today can clean up its own dirty nest and ignore the rest of the world. California spent billions of dollars curbing pollution levels only to confront emissions from Asia undermining its clean-sky policies. The scale of our impact calls for a radical overhaul of our industrial age technologies, economies, and governance, and the need to bridge the deep schisms between people and nations caused by globalization.

Over the years my views have changed from seeing conservation as a bulwark against development and development as an obstacle to conservation to seeing the two as the yin and yang of our aspirations. Moving up the ladder of concern from conservation for survival to conservation for planetary health depends on ecological emancipation as well as economic and social development.

A seminal contribution of conservation science has been the recognition that open space, mobility, and complex ecological processes underpin the diversity, productivity, and resilience of ecosystems.[4] All three are crucial to sustaining wildlife, habitats, biodiversity, and the natural process supporting all life and our own well-being.

Putting a premium on space underscores the need for good conservation practices across the other 83 percent of the land surface no less than in the 17 percent within parks. Setting aside protected areas is not enough. We can't build Berlin Walls around nature and expect to preserve it intact or restore it to a Pleistocene facsimile, given the changes we have created.[5]

I should lament the loss of the vast wild spaces of southern Tanzania in the 1950s, and the traditional Maasai culture of Lakato's time. I don't. In hindsight I valued the remote reaches because I was insulated from the hardships and hazards of making a living off the land, raising a family, and living to old age. The world of my upbringing has vanished. Today I enjoy the freedom to savor wildlife and wild places and to explore breathtaking landscapes, exciting cities, and exotic cultures around the world more than I ever did on the narrow bush trails of my youth. I delight in seeing a young generation of Kenyans and overseas visitors enjoy wildlife and, frankly, I get greater pleasure helping them do so than I would living alone by my Walden Pond.

The world has changed irrevocably with globalization. In Kenya I worry about the communities left behind in the rush for development, rocked by political conflicts and displaced by land grabs. I fret over the destruction of their land by commercial trucks hauling out charcoal, sand, and rock as well as by the erosion of topsoil, the dwindling grasslands, poaching, and the shrinking spaces left for pastoralists and wildlife. Yet Kenya, like Wallace Stegner's American West, is a land of eternal hope.

In the 1980s Wes Henry and I calculated that a lion in Amboseli was worth $10,000 a year—equivalent to a hundred cows at the time—if the Maasai were to cash in on tourism rather than hunt down lions and sport their shaggy manes as headdresses. The Maasai didn't get anything out of tourism for a long while, despite our government policies to engage communities in conservation, and wardens did little about lions killing their livestock. Lion numbers fell from fifty to ten in the ensuing spearing and poisonings and recovered only when communities set up wildlife and tourism enterprises and conservation organizations like Big Life Foundation and Lion Guardians paid Maasai scouts to protect lions and curb conflict. The South Rift Association of Land Owners, assisted by the NGO Rebuilding the Pride, has built up the lion population from ten to eighty since setting up its own wildlife conservancy along the Kenya-Tanzania border remote from any park.

These are pinpoints of light. But how can we conserve the land-hungry endangered lion, elephant, and grizzly in a populous world, much less the great whales plying the world's oceans beyond the reach of national laws and jurisdiction? The global commons are our final frontier. Either we clean up the dirty nest we are making of it, or we stew in the mess of our own making.

Stewart Brand epitomizes a shift from the environmental pessimist who founded the *Whole Earth Catalog* in the 1960s to an unapologetic optimist in his latest book, *Whole Earth Discipline*.[6] Brand now advocates new technologies rather than population and economic shrinkage as the solution to planetary disruption. We can use satellites and sensors of all kinds to monitor the state of the planet, he says. We can blunt climate change using geoengineering skills such as seeding the oceans with iron to boost algal production and, along with satellite-launched umbrellas to shade the Earth from the sun's rays, mop up excess carbon.

I see no need for geoengineering to reduce greenhouse gases—except as a desperate last-ditch effort should all else fail. Brand himself gives reason for caution: most billion-dollar programs have failed, he says, even as he talks up trillion-dollar quick planetary fixes. I share the public's wariness of technological fixes. Technology is moving too fast for the general public to grasp, according to a survey of twenty-seven countries designed to measure confidence in the public and private sectors.[7] Governments and businesses give too little thought to the impacts of technology, whether the proliferation of social media, digital security, genetically modified foods, or fracking, and how unevenly the benefits are shared.

Scientists Anthony Barnosky and Elizabeth Hadly highlight the perils of technological change without first understanding the pros and cons of new innovations.[8] With family size shrinking and rich societies reaching a peak of materialism, rediscovering satisfaction in experience rather than things, a growing public demand for a cleaner world and greener products may in any event shape the products of technology. British economist Nicholas Stern agrees: the tide can turn quickly when technology, economics, morality, and politics align, as the pushback against smoking, drunk driving, and HIV show.[9] Environmental journalist Fred Pearce also sees green shoots of hope.[10]

In the last resort, high tech could be the only thing that can save us. As a thought experiment, imagine glaciers advancing halfway down America and Europe in a new ice age. The cool, dry climate would compress arable lands, slash food production, shrink tropical forests to a tenth of their current size, and threaten biodiversity more gravely than the two- to four-degree-centigrade rise in global temperatures projected by the Intergovernmental Panel on Climate Change (IPCC). Would we let nature take its course or try to turn back a new ice age? We may unwittingly have done so already. The conversion of vast tracts of forest and woodlands to farmlands in Asia and Europe seven thousand to eight thousand years ago vented enough carbon dioxide into the atmosphere to check an ice age in the making. Our ancestors may have done us a favor in avoiding a big freeze, yet they set the stage for a hot Earth.

The fossil fuel emissions of the industrial age have built up an environmental debt that makes each ton of new pollution more damaging than the last: a ton of carbon vented in the nineteenth century was easily absorbed by the natural capacity of the vegetation, soils, and oceans. Today, each ton adds to a large base level, elevating atmospheric CO_2 by over a third more than pre-industrial levels, reducing the natural buffering capacity of the Earth, and driving up the cost of holding temperature within the two-degree-centigrade target of the Paris Agreement. China and India don't have the luxury of starting from the near-zero emissions of the West at the dawn of industrialization or of kicking the cleanup costs decades down the line. The black soot and urban smog enveloping northern India in late 2017 rose to 150 times the safe limits, causing millions of respiratory complications, dozens of deaths, and a drag on the economy.

Pollution has shifted from a relatively few localized industrial centers during the nineteenth century to millions of small enterprises and billions of households in the post-industrial economies of the twenty-first century, further complicating remedial action by governments and free-market solutions.[11]

Reconnecting the proximity effects that make us responsive to our actions is, in my view, the surest way to unlock the logjam to cleaning up our dirty global nest. Erematere is a useful metaphor for reinstating the severed links connecting our actions to their consequences. I say metaphor because the natural connections in the erematere stewardship in small, close-knit biocultures can't work on a global scale. The natural connections have been so ruptured by economies of scale and the cornucopia of goods we buy at bargain retail prices in supermarkets that we lose sight of the external and future costs of offshore factories spewing pollutants, toxic chemicals, and greenhouse gases into rivers, air, and oceans.

The tragedy of the global commons is the greatest of all the challenges in our Human Age. Trawlers from Europe, North America, and Asia have depleted fish stocks from the Arctic to Antarctic. Worst hit are the large, slow-growing species like whales and sharks, and shoaling species like cod. The threatened bluefin tuna, a delicacy in Japan, fetches up to $450 a kilo. In 1994, as director of the Kenya Wildlife Service, I was asked by the World Wildlife Fund to have Kenya, which enjoyed good relations with Japan, submit a proposal to list the bluefin tuna as an endangered species at the upcoming meeting of CITES (the Convention on International Trade in Endangered Species). The submission was as scientifically sound as the case for declaring the elephant an endangered species and securing a worldwide ban on ivory trading in 1989. All the same, the day after filing the submission I was called to the minister of wildlife and tourism's office: the Japanese ambassador has got wind of the submission, he told me. Either you withdraw the CITES proposal or Kenya loses Japan's foreign aid. The bluefin numbers continued to plummet.

At the CITES convention held in Fort Lauderdale, Florida, where the bluefin proposal was to be discussed, listing a species as endangered or threatened was less a matter of scientific merit than political deals, arm-twisting, and bribery by wildlife traders taking national delegates out for a quiet meal. Later, at the 1997 CITES meeting held in Harare, Zimbabwe, I headed up the Kenya delegation once more. This time, in a closed session of African states free of backdoor political deals and bribery, we agreed on mechanisms for rating the degree of threat to the elephant based on scientific criteria and continual monitoring. Progress can be made in the global commons, provided the transparency, scientific scrutiny, regulations, monitoring, and enforcement that works locally are in place.

The greatest challenge of the global commons lies in the remoteness, invisibility, and intangibility of our global ecological imprint. How can we engage our senses, minds, emotions, and social sensibilities if we can't see, feel, track, imagine, and project how our actions will affect both us and the health of the planet?

Ecology in its literal Greek translation means the science of the home. We should view the planet as our communal home and ecology as our domestic science. We keep our own home clean and wash our hands to avoid diseases and contamination. Our neighbors yell at us and turf our trash back in our yard if we dump it in the street. So why are we blind and uncaring about our Earth's polluted rivers, oceans, and skies, and what can we do about it?

What we can do is use unnatural technological surrogates to bring the planetary ills within the orbit of our domestic concerns, where proximity and feedback bind together communities and neighborhoods. The World Wide Web opens a way to reconnect the proximity effect and erematere linkages of close-knit communities severed by globalization. Humans throughout history have used new technologies and social forums to re-connect their fragmented activities in expanding from local communities to a global marketplace.[12] The innovations include the steam engine, car, plane, telegraph, radio, telephone, and internet, exchangeable currencies, laws, regulations—and international agreements such as the World Trade Organization, the Paris Climate Accord, and CITES. Each innovation enables us to connect over ever-larger distances at greater speeds. The open networks of the World Wide Web scale up social connections from the few hundred yards of the medieval town crier to the global reach of the Z Generation.[13]

Communications specialist Howard Rheingold, in *Smart Mobs: The Next Social Revolution*, attaches great importance to trust and reputation in the World Wide Web.[14] He sees the Public Goods game noted in chapter 9 as forging a remote reputation system among social and professional users of the internet. The propensity of small, close-knit communities to build up social capital is resurfacing in the internet community, where users show greater generosity than unbridled self-interest dictates, and penalize cheats in the interests of the group. The Web creates a network of individuals and a free flow of information that level the playing field of privileged access to information and opinion making.

In short, rules governing reciprocity among the Maasai and the fuldo trading networks of the Konso have reemerged time and again in our global age via Facebook, WhatsApp, e-Bay, Uber, and Airbnb. Social networking has driven the technological revolution in personal computers and cell phones by expanding the natural senses of our ears and eyes to engage each other in a personal way around the world.

Enormous hurdles stand in the way of using the Web as a surrogate of our natural senses and collaborative nature, though.[15] The openness of the Web poses the same political threats facing all open democracies and societies—propaganda, manipulation, and marginalization—but on a far larger scale, making these perils harder than ever to combat. Russia's meddling in the 2016 U.S. election bears witness to the undermining of the most basic tenet of democracy: free and fair elections.

Another challenge of the World Wide Web is the flood of information. How can we possibly search, absorb, and distill down the yottabytes ($1,000^8$ units of information) crowding cyberspace? Anticipating the challenge,

Vannevar Bush, head of the U.S. scientific effort during World War II, urged that we expand our minds using machines able to manage and distill the flurry of information so that we can grasp and use it effectively.[16] Seventy-five years later, Wikipedia, Google, Amazon, and other vendors were deploying search engines and algorithms to mine and analyze vast quantities of data and tailor books, films, and information to suit our needs and tastes. In the making are digital libraries of every article and book ever published, linked to search engines and protocols able to access any information instantaneously.

Here again, we need to be cautious in treating the past as prologue for the future. Communications specialist Andy Clark warns us of typifying human nature by what it used to be. Human brains are so plastic they enabled us to move from communication through our natural senses to communication via computers, the World Wide Web, and smartphones in half a generation. Extrasensory technologies using numbers, alphabets, and images are, for Clark, "mindware upgrades."[17] Shortcomings aside, the new technologies geared to our natural socialness and the inclusivity of small, close-knit communities are ushering in the fourth industrial revolution.[18]

The first industrial revolution used water and steam to power production; the second used electricity in the service of mass production; the third used advances in electronics and information technology (IT) to automate production. The fourth industrial revolution differs from its predecessors in scale, scope, speed, complexity, and impact, and is evolving at an exponential rather than linear pace. The exponential speed and global scale are disrupting every industry as well as the way we produce goods and services and manage our lives. Artificial intelligence (AI), robotics, autonomous vehicles, 3-D and 4-D printers, nanotechnology, biotechnology, energy storage, and quantum computing capacity are connecting people, information, and peripherals of all sorts, creating the Internet of Everything (IoE).[19]

The fourth industrial revolution is changing the nature of business too, by favoring nimble and innovative start-ups with low entry costs to the marketplace. The innovations have shrunk the average longevity of businesses from eighteen years to eleven in response to consumers demanding greater transparency and, through the power of social networks, forcing companies and service providers to adapt designs, marketing, delivery, and services more suitable to their individual needs.[20] The public is also demanding a bigger role in political decision making and policies by pushing governments to be more transparent and, as with businesses, to be more agile, efficient, and inclusive. In this sense, the fourth industrial revolution is being shaped by local communities and across borders rather than by big governments making decisions within nation-states, as in the nineteenth century.[21]

New developments in technology suggest that innovations in robotics and AI will further transform everyday life, employment, and leisure. Robots could reverse offshore jobbing by making cars, computers, dishwashers, and TVs as cheaply in the West as in India, China, and Bangladesh.[22] As Mark Weiser of Xerox points out, most technologies, such as heaters to warm us up and sensors to detect lighting levels in our houses, will become tiny nano-machines and sensors embedded in our clothing, around the house, and in our cars, making them virtually invisible and an integral extension of our body through the IoE.[23] Some 13.5 billion devices were connected to the internet in 2015. By 2020 the number will top 50 billion, including those embedded in parking spaces, streetlights, garbage cans, and just about everything else we use. Some fifteen thousand or more new apps are being developed weekly. The biggest gains in efficiency and waste reduction, as I will detail in chapter 21, are made in the cities, which account for 55 percent of the world's population and 70 percent of the annual greenhouse gases emissions. Sensors planted strategically in our cities and around the biosphere can detect and monitor the impact of our actions in real time and expose the abuses of governments, corporations, cheats, and free riders.

The fourth industrial revolution opens another novel frontier, designer genes—the modification of genes and organisms for specific purposes. Geneticist and entrepreneur Craig Venter created the first artificial life, the phi X174 virus, using computer information to code DNA. He then inserted the manufactured DNA in a bacterium to create an organism that fed, moved, metabolized, and replicated itself.[24] Venter believes synthetic biology has the potential to transform every aspect of life.

The biosynthesis of species raises the question of how we use science and technology, whether for public good or private evil. Synthetic biology opens the door to replacing deleterious genes in critically endangered species and creating bacteria able to mop up wastes, denature toxins, absorb greenhouse gases, and convert CO_2 to carbon compounds able to enrich soils. On the downside, we risk creating Frankenstein monsters in the genomes of domestic crops and livestock at risk of escaping and disrupting ecosystems. The prospects raise a host of ethical and practical dilemmas—whether, for example, we should resurrect extinct species such as the dodo, moa, passenger pigeon, thylacine, and wooly mammoth, or manufacture entirely new species such as bacteria to digest oil slicks. How do we arrive at socially acceptable, ethical, and environmentally sound decisions?

Columbia University's Center for Research on Environmental Decisions is studying the processes shaping our everyday choices, behavior, and attitudes.[25] Drawing on economists, psychologists, and anthropologists from around the world, the center is looking into environmental decision making—how, for example, our perceptions of risk and uncertainty shape our responses to climate change and other threats. The work confirms the findings of Nobel Prize winner Daniel Kahneman and associate Amos Tversky: confronted with choices, people have automatic biases. For instance, we are more averse to losses than gains, make repeated errors in judgment because of our tendency to use shorthand rules to solve problems, and opt for smaller rewards now rather than bigger ones later.

Cognitive psychologists widely agree that we use different systems for processing risk. One is analytical, based on sizing up information; the other is based on experience, ingrained feelings, and quick emotional responses. The first undervalues delayed benefits; the second underestimates the risks of remote threats we haven't experienced—sea-level rise, for instance. Urbanites in Alaska see climate threat as far away and down the line and the solution as switching on an air conditioner occasionally in summer. Rural farmers, ranchers, and fishing communities see the threat up close and personal in the hotter weather as well as more intense fires, floods, and droughts threatening their livelihoods.

The failure to convince climate change skeptics is often seen as a failure of communication skills in scientists coupled with the short-termism of politicians loath to address problems beyond their reelection cycle. Ninety-eight percent of U.S. federal funding on climate change goes to physical and natural sciences research, but far more needs to be spent on the questions of how public perceptions are formed and how public responses are made to threats.

Columbia's Center for Research on Environmental Decisions is also investigating public decision making. Researchers find it easier to forge cooperation if members join the team from the start rather than deliberate individually first. Group decisions encourage "we" and "us" talk more than "I" talk, creating a good feeling of being part of a community and greater commitment to its outcome.[26]

Richard Thaler (who won a Nobel Prize for his pioneering economics work) and Cass Sunstein, in their influential book *Nudge: Improving Decisions about Health, Wealth, and Happiness,* show that decision science, based on how we think in real life rather than neoclassical economic models, can help us confront and address threats.[27] We do better at cutting electricity use with constant visible feedback from home meters than from government calls for energy savings. Imagine, then, the reduction in home energy and water consumption if every household appeared on Google Map, highlighted in green for responsible behavior, yellow for marginal, and red for reckless.

Reconnecting the social animal in us at a global scale holds out green shoots of hope, not only for harnessing the fourth industrial revolution to fit our needs and aspirations, but also in recapitulating the conditions that engage cooperative capacity, foresight, and responsibility for our actions.[28] How can we tackle a range of challenges, from individual survival to planetary health, by using our natural connections amplified by extrasomatic surrogates?

17

New Tools for a New Age

Fifty years after starting my study of Amboseli, I'm still torn between optimism and pessimism. On the bright side, the elephant herd has grown threefold and spread far beyond the ecosystem since poachers gunned down two-thirds of the population in the 1970s. The herds are safe today, protected by 375 community scouts. This is a rare elephant success story at a time of resurgent ivory poaching fueled by thriving markets in China and East Asia. On the downside, the very success of community-based conservation around the Amboseli National Park has caused a new and more troubling threat—conflict with farmers.

Wildlife is being squeezed, harassed, and killed as the open savannas are hemmed in by a welter of farms, ranches, and sprawling suburban settlements. Elephants, drawn to maize and bean farms south and east of the park, have killed several people defending their crops and children walking to school. Riled at Kenya Wildlife Service's rapid response to poachers but failure to deal with crop-raiding elephants, gangs of warriors spear the first elephant they come on in reprisal, upping the chances of further attacks by stressed-out elephants.

In reflective moments I take consolation in the arrow of progress pointing upward. Paul Ehrlich's predictions in *The Population Bomb* of massive famine, starvation, and collapse in Asia by the 1980s proved wrong. Agricultural productivity rose by two-thirds in Asia between 1960 and 1980, and twofold worldwide between 1970 and 1990, driven by high-yielding varieties of rice, maize, and wheat. Worldwide, famine is receding, absolute poverty has shrunk from 25 percent to 13 percent over the last quarter century, incomes are rising, the gap between rich and poor nations is closing, and population growth rates have fallen sharply.[1]

Despite the positive trends, Ehrlich still sees little hope for humanity or nature. In 2015, Ehrlich and Corey Bradshaw compared pre- and post-industrial impacts on the environment in Australia and America.[2] Among other factors, they blame the environmental degradation on politics, greed, ecological ignorance, and a blind trust in technology to solve the problems. The solution, they argue, is scaling back to a population of 1 billion and freezing economic growth. This is not a view I share, and neither do the likes of Jonathan Lebo, bent on enjoying the comforts and freedoms of the West.

Matt Ridley gives another view in *The Evolution of Everything: How New Ideas Emerge*.[3] Most great leaps in human development happen from the bottom up, not by design, he argues. I share his view and have championed bottom-up solutions to conserving species and ecosystems. I don't share his faith in free-market solutions, though, any more than I subscribe to Bradshaw and Ehrlich's solution of extreme population reduction.

Computer projections based on massive data sets and rigorous calibrations show a more nuanced and sanguine outlook for the population than the dire projections of the Club of Rome's *Limits to Growth* in the 1970s. Joel Cohen, in *How Many People Can the Earth Support?* finds no fixed upper limit for human population or tipping point for resource depletion.[4] Rather, the Earth's carrying capacity depends on how we manage our resources and the quality of life we aspire to.

In *Feeding the World: A Challenge for the Twenty-First Century,* Vaclav Smil, distinguished professor of geography at the University of Manitoba and rigorous analyst of world food sufficiency, concludes that current agricultural outputs can feed 10 billion people if, and a big *if* it is, we are guided by equity in consumption and eat a frugal, largely vegetarian, and nutritionally adequate diet.[5] Repeat the American levels of consumption, waste, inefficiencies in production, and environmental impact, and the world can't feed even 6 billion people for long. For Smil, food is not the limiting factor. With population slowing for other reasons, we should focus on producing a healthy diet without damaging the environment and improving efficiencies along the entire food chain, from crop production to storage, delivery, consumption, and waste disposal.

Despite his conclusions, Cohen is not optimistic about dispelling the myth of overpopulation, and with good reason. Psychological studies show that the very act of dispelling a myth tends to reinforce strongly held beliefs. Hans Rosling and colleagues have shown how biases consistently lead to underestimates of human progress and development indicators such as a reduction in fertility and poverty and increases in women's education.[6]

On the plus side, the world has reached a political consensus on the principles of sustainable development adopted by the World Commission on Environment and Development in 1987. In 2000 the United Nations pledged to meet a set of common goals by 2015. Dubbed the Millennium Development Goals, the targets included the eradication of extreme poverty and hunger; universal primary education; gender equality; empowering women; reducing child mortality; improving maternal health; combating AIDS, malaria, and other infectious diseases; ensuring environmental sustainability; and setting up global partnerships for development.

Many of the results fell far short of success in 2015, mainly in the poorer African countries and in the case of environmental quality. The links between conservation and development were far too weak.[7] Too weak, for example, to deter Mukogodo herders from chopping down trees for charcoal to feed their families today in order to have any hope of a tomorrow. The linkages were made far more explicit in the 2015 United Nations Sustainable Development Goals. Supply the clean water, food, and energy needs of poor families and the future lengthens and with it, the prospects for conservation and sustainable use of resources.

Of all our basic needs, fresh water is in shortest supply. The amount of water on our planet has stayed constant since the formation of the Earth 4.4 billion years ago.[8] The implications for conservation are profound: unlike forests and soils, we can't destroy water, but we can waste and pollute it, run short for lack of planning and infrastructure—and go to war over the rights of access and offtake. This makes water a good starting point for exploring the principles underlying the use of scarce resources in general, simply because we can't manufacture our way out of water scarcity, as we can a shortage of fertilizer, or use substitutes, as in the case of wood and oil supplies.

In *The Water Kingdom: A Secret History of China,* Philip Ball recounts how water has shaped the rise and fall of empires.[9] Unmanaged, too much water flooded and drowned communities; too little and it brought on droughts and famines. China's rise to a powerful centralized state two thousand years ago stemmed in part from building the first dams to regulate the flow and provide a year-round supply of water for crops and urban populations. The scourge of dysentery, cholera, and other water-borne disease brought down empires and still causes thousands of deaths in poorer and remoter regions in the aftermath of floods, earthquakes, and other natural disasters.

We either control, manage, and allocate clean, safe water supplies equitably or risk wars and disruption. In 2018 Cape Town was predicted to be the first major city to run out of water and narrowly escaped Day Zero by rationing and cutting water use in half. Twenty-one major cities in India face similar water crises, a third of all households have no sanitation, and water-borne diseases account for a third of all deaths.

Water crises can be avoided with long-term conservation planning in efficient use. Fishman notes that the volume of water used in the United States has fallen from a peak of 1.7 trillion liters in 1980 to 798 billion in 2005, staving off a water crisis, despite the addition of 70 million more people and GDP doubling from $6 to $13 trillion in the same period.[10] Efficiency gains stemmed from better storage and delivery facilities in power plants and on farms. Farms, for instance, used 15 percent less water in 2005 than in 1980 for a harvest half again as large.

No place illustrates the case for water conservation better than the American West. In *Water Is for Fighting Over: And Other Myths about Water in the West*, John Fleck says Mark Twain never made the famous comment about fighting over water—but it remains too good not to use anyway.[11] Fight over water and everyone loses. Water use need not be a zero-sum game in which one person's gain is another's loss—as the Bedouins showed in constructing elaborate underground water channels as the Middle East climate dried. Water conservation is a win-win proposition if managed well.

Water is a local rather than a global problem. Conserving and managing water depends on cooperation among farmers, homeowners, and industry, and between cities, states, and countries. The scarcer the water supply, the greater the need for thrift, high-level arbitration in its use and distribution, and innovative technical and economic solutions.

A good example is Las Vegas, the American capital of entertainment plunked in the Nevada desert and famous for its opulent hotels and water fountains.[12] Patricia Mulroy, head of metropolitan water, showed how water shortages can be offset. Taking a comprehensive view of water supplies and planning carefully, Mulroy shrank water use per person from 1,290 liters to 921 liters between 1989 and 2009, at a time of strong economic growth. She persuaded the casinos and hotels to restrict use to recycled water off grid, upped water rates, introduced progressive pricing (a rise in the cost per gallon in proportion to the volume used), amalgamated water districts, and increased availability through regulated distribution. Golf courses were set a ceiling of freshwater use, forcing them to use recycled water. Revenues from the increased water rates were pumped into extra storage capacity and, through a public relations campaign, into winning community support for a culture of water thrift as a way of life.

California, the biggest state economy in the United States and the thirstiest, produces over half of America's fruit and nuts. The irrigation of water-sapping crops like tomatoes, oranges, and almonds accounts for 80 percent of the state's consumption. A tradition of individual water rights has also made California one of the most profligate water states in the country. Weak regulations have resulted in water tables dropping up to fifteen meters in three decades in the Central Valley, buckling roads and cracking bridges. Wells are running dry and farmers are hunting for new water supplies. A seven-year drought, which finally broke in 2017, and climate warming have added to California's water woes. Winter snowfall has dropped and spring runoff has risen, making it harder for city and state authorities to capture and store runoff, and water has become more expensive as a result.

If California becomes as water stressed as Australia—and that could soon happen, according to current climatic projections—the state will need to introduce measures to combat persistent drought. The steps Australia took during the Big Dry, a crippling twelve-year period between the late 1990s and 2000s, point the way forward. Water rights were rescinded and reallocated, storage and delivery facilities were boosted, and efficiency was raised by astute pricing policies. Melbourne's per capita water use dropped a staggering 60 percent, from 360 to 144 liters per day.

In *The Big Thirst*, Fishman makes a strong case for the benefits of efficiency in water conservation by showing that businesses, power companies, and agriculture in the United States have all made big strides in conservation and water-use efficiency—because it pays. IBM trimmed water use 29 percent between 2000 and 2009, saving $740,000 in annual water bills. Big Blue now uses five thousand sensors along its production lines and analyzes 400 million data points to improve feedback and efficiencies. For every $1 million cut in its water bill, IBM saves $4 million in chemical, electrical, and energy costs. Adopting Roosevelt's "Gospel of Efficiency" and keen to tout corporate social responsibility, IBM, Coca Cola, Intel, and GE are posting water usage online. If all residential, business, and government users adopted similar voluntary initiatives, imagine the outlook for water supplies. The role of water agencies could be scaled back to ensuring cooperation among users, saving the taxpayer money, and reinvesting the savings in long-term planning, monitoring, and regulation.

The path to water use efficiency boils down to pricing the true cost of storing, cleaning, and delivering water, rather than treating it as a free public good, as if it were the air we breathe. In the United States the average household pays $30 to $50 a month for water. The price is so low that unlike gas prices, a several-fold raise barely reduces consumption. Treat water as if it were 90 percent of our own body, which it is, and we would look after it more carefully. In East Africa's drylands, where water is sparse, the Swahili saying goes that *maji ni maisha,* water is life, because households carrying four-gallon loads from rivers and wells two miles away treat it as lifegiving as well as lifesaving.

Like water, food production and waste containment stand to gain quick wins from good husbandry and conservation practices. The agroindustry production system, warding off mass starvation as the world population grew from 1.6 to 6 billion in the twentieth century, succeeded in boosting crop yields through the development of new varieties of high-yielding super crops, such as rice, maize, wheat, and soybeans. The hitch in the super crops of the Green Revolution is the incremental use of nitrogen (N), phosphorus (P), and potassium (K) fertilizers and pesticides needed to achieve high yields. Food production will need to climb a further two- to threefold further to meet peak population demands and the minimum calorific requirements of 11 million people by the end of the twenty-first century. Achieving peak food sufficiency using twentieth-century agroindustry technology, fertilizer, and pesticide methods is unachievable and would in any event cause irreparable damage to soils, rivers, lakes, and oceans, spark acute water shortages, and increase green-house gas emissions and global warming, which will in turn depress crop yields and increase rather than cut agricultural water use. How can the lockstep increments in food production and damage to the environment and health be broken?

The quickest and cheapest way is to boost fertilizer efficiency and cut waste. Based on the historical use of nitrogen in 113 countries over the last half century, Xin Zhang and colleagues found that applications first grew then shrank as a function of GDP.[13] The transition from high to low use can be measured by nitrogen use efficiency (NUE). EU countries uncoupled fertilizer use and yields by slashing subsidies and application rates through Common Agricultural Policy directives. Xin Zhang and colleagues project that world food needs can be met within the current 1.3 billion hectares of farmlands and safe environmental limits if—and again, a big *if*—we reduce the global nitrogen surplus by half to the levels currently achieved with best practices in the United States and France.

The history of every nation is eventually written in the way in which it cares for its soil. Such was the message President Franklin Delano Roosevelt put out to America during the Dust Bowl era, laying the foundation of farmland conservation. Soil is the undercoat of the planet and it, along with the atmospheric topcoat, is pivotal to planetary health in the twenty-first century. Improvements in the efficiency of fertilizer use at the farm gate depend on the 4Rs: the right fertilizer used at the right rate at the right time in the right place. Advances turn on better husbandry practices rather than brute machine power and ad lib fertilizer use.

Further advances will depend on new technologies able to release fertilizers and deliver water and nutrients just in time for direct root uptake. Efficiency improvements can also be attained by high-tech sensing and monitoring tools, computer analysis, and the sort of modeling and fore-casting IBM uses along its production line. Another example is extraction and reuse of carbon, nitrogen, and phosphorus from farm wastewater to cut import costs, save energy, and boost profits. Such integrated crop management based on ecological principles of land and food husbandry is a time-tested way of using biocultures to boost crop production, lower losses from pest infestations, and reduce environmental damage.

Several conservation biologists argue that such intensified agricultural techniques and improved efficiencies, because they free up land for biodiversity, are virtuous, producing higher crop yields from less land.[14] Between 1965 and 2004, agricultural intensification saved 27 million hectares of land from being cleared for farming and, for the first time since the dawn of agriculture, took more land out of cultivation than went into it—an area amounting to half the size of the UK each year.[15] I share the land-sparing view, having watched poor farming practices in Kenya lower yields to a tenth of those obtained in California and trap farmers in a downward spiral of eroding soils, declining water capture, failing harvests, and declining livestock yields. The downward spiral spells Mukogodo-like poverty and no place for natural habitats or wildlife.

The wild cultivars we domesticated as cereals, fruits, and vegetables in the Neolithic didn't evolve to suit us. Most were small, often bitter, and so low in calories that it took early human gathering parties, like baboons, hours to harvest enough to fill a stomach. Farmers and herders using Darwin's artificial selection took thousands of years to boost the calorific content of cereal crops, fruits, and vegetables and the growth rates and fat content of livestock. The pace of genetic modification jumped sharply in the twentieth century with the use of genetic zapping techniques such as X-rays, toxic chemical mutagens like colchicine, and cobalt and atomic radiation.[16] Aside from the inherent dangers of lethal substances, the sledgehammer techniques sacrificed genetic diversity for higher yields.

Down the line, the genetic revolution holds promise of producing more nutritious and less fattening produce friendlier to the environment than current breeding techniques. Biologists are widely agreed on the prospects for genetically modified organisms and crops, seeing scope for new techniques to boost yields substantially.[17] Based on bacterial countermeasures to viral invasions, individual genes can be tailored by CRISPR Cas9 and Cas9HF techniques, speeding up the selection of favorable plant traits to a matter of weeks rather than years—without reducing genetic diversity. The CRISPR techniques make it possible to snip out, insert, replace, or silence genes, opening the door to highly targeted alterations that adapt crops to local environmental conditions.[18] So, for example, genes targeting specific loci on plants and sprayed onto fields could boost yields, nutrient content, and resilience to heat, frost, and drought. Sprays able to kill pests by disabling viral genes rather than broadside application of heavy pesticides are already in development.[19]

Energy presents a similar situation. The quickest wins in sustainable and containable use of energy can be made using conservation principles. For decades in the twentieth century, GDP worldwide was tightly coupled to energy consumption: every increment in the economy meant a proportionate increase in use of energy and disposal of waste. Cuts in oil production imposed by the OPEC cartel in the 1970s forced up fuel prices, increased fuel efficiency through improvements in engine design, and uncoupled the GDP-energy-pollution linkage. Energy efficiency (most notably in Denmark, in Europe generally, and to a lesser extent in the United States) has continued to advance, despite a resurgence of oil output and price slumps.

The message is clear: energy shortages and price hikes, like droughts and famines, change our mindset and gear us up to anticipate, learn, and ultimately refashion society to prevent and mitigate disasters. National security concerns over dependence in the United States and Europe on volatile Middle East supplies in the 1980s, projections of oil supplies peaking in the early twenty-first century, and the impact of greenhouse gases on global climate in the 2000s have changed the outlook for the future of energy in two main ways. First, policy and markets have shifted toward renewable energy, which widely distributes the sources of production and zeroes carbon emissions. Second, technical innovations are opening vast reserves of oil shale to offset the "peak oil" crisis. Fracking has temporarily averted an energy-induced economic slowdown, although it has also slowed a transition to renewable energy.

Another factor stalling the uncoupling of economic growth, energy use, and pollution emissions is the pretax subsidies to the oil industry, totaling $550 billion worldwide.[20] The IMF points out that the subsidies ignore the deleterious effects of fossil fuels on health and the environment. A 2016 report by the World Bank puts the price at 6.5 million premature deaths and $5 trillion in total welfare costs, counting the lost work time and lowered productivity worldwide.[21] Fossil fuel subsidies discourage investment in renewables, which have no hidden production costs. Renewables also liberalize energy production through the household generation of solar and wind energy. The economic mispricing of fossil fuel is the biggest hurdle to the development of clean and sustainable energy.

Removing the subsidies, using nuclear power as a bridging source low on greenhouse gas emissions, and pricing the health and the environment costs would uncouple the GDP-energy-pollution linkage and speed up the adoption of renewables out of economic expediency.

Is the potential for renewable energy adequate to supply present and future populations and economic growth? Wind, solar, wave, geothermal, and hydroelectric power sources combined can meet most of the present and projected demands. In 2015, for the first time, renewables accounted for the bulk of new electricity generating capacity worldwide, with China, India, Brazil, and other emerging markets accounting for over half the $286 billion invested in wind, solar, and other renewables.[22] Excluding hydro, 10.3 percent of all electricity generated worldwide came from renewables, twice the 2007 level. The average global cost of solar dropped 61 percent between 2009 and 2015 and by 2020 had become cheaper than coal. Germany has already achieved 30 percent of its electricity from renewables and aims to achieve 100 percent by 2035. Kenya plans to convert to a post-oil green economy by 2030.

Improvements in water, food, and energy uses are one part of the sufficiency and sustainability equation. Another is breaking the industrial revolution paradigm of seeing bigger technological equipment and factories as the engines of economic growth and prosperity. How can that be done?

I noted in the last chapter how the fourth industrial revolution opens prospects of harnessing technology to fit human needs and reconnecting the social and ecological linkages severed by the global marketplace and agroindustry economies. The fourth industrial revolution must go hand in hand with another change needed to break the link between growth, consumption, and destruction—a shift from the linear economy of the twentieth century to the circular economy of the twenty-first century.[23] The circular economy is about refashioning and reusing materials, increasing production efficiency, and reducing waste. In this sense the circular economy mimics the energy flow and nutrient cycling of ecosystems, scaled up to a planetary level. Progress toward a circular economy can be gauged by the efficiency of resource use—the amount of material used to generate a dollar of GDP. So, for example, China presently uses 2.5 kilos of material to generate $1 of GDP, while the Organisation for Economic Co-operation and Development (OECD) countries, such as Britain, France, Germany, and Japan, use 0.54 kilos for every $1 of GDP. Resource efficiency worldwide rose by 34.7 percent between 2005 and 2013.

Australia is one of the few countries to assemble the skills needed to take an integrated look at how to achieve the twin goals of sustainable growth and a healthy environment. Steve Hatfield-Dodds and seventeen other researchers at the Commonwealth Scientific and Industrial Research Organisation (CSIRO) ran a series of models using a variety of assumptions on how both growth and sustainability can be met, focusing on agriculture, energy, and transportation.[24] The models show that the Australian economy and living standards can grow strongly to 2050, and that carbon emissions can be reined in from the present level four times the global average to matching levels. One-third to one-half of the decline in carbon emissions can be met by sequestering carbon through reforestation and habitat restoration, which will simultaneously boost biodiversity protection and cut extinction risks by 7–9 percent. The rest of the emission reductions would come from increased energy efficiency. Water use, slated to double by 2050, could be met by desalinization in coastal cities and recycling for industrial use, for instance. Most of the uncoupling of economic growth and environmental depletion can be achieved through existing policies and by mobilizing efficient technologies and use of resources. All it will take is a strong commitment by Australians.

The United Nations has made huge strides in reducing the incidence and impact of wars, contagious diseases, and disasters, but still lacks the monitoring and enforcement elements for solving global tragedies of the commons such as the COVID-19 pandemic, global warming, and acidifying oceans. If we can't make sound decisions to clean up our dirty nest for the sake of our own health, we certainly won't do so for the sake of the birds and bees. If we do make the connection between our own welfare and well-being and the environment, the prospects brighten for all life. How can we clean up our dirty nest for our own sake and create a livable world for the birds and bees?

18

Cleaning Our Planetary Nest

Nature is a veritable arms race in which prey must counter every new predator adaptation. Millions of years of evolutionary predator-prey parry and riposte show up in the toxic repellents and flamboyant warning colors of butterflies, nudibranch mollusks, and gaudy poisonous dart frogs. Palatable species take advantage of warning coloration by mimicking poisonous species. The North America viceroy butterfly, for example, mimics the coloration of the more poisonous and common monarch butterfly.

We have our own inbuilt evolutionary revulsion of snakes, spiders, and creepy crawly creatures inherited from our ancestors. Like all primates, we are also persnickety and cautious when it comes to unfamiliar foods that might poison or sicken us.

I learned about our ancestral caution trying out new foods at school in the southern highlands of Tanzania. A small gang of bush mates—from whom I drew my nickname, Jonah—and I explored the nearby hills every Sunday. As ever-hungry teenagers, we were always on the hunt for wild foods. Berries and fruits were plentiful, but many were spitting-bitter and, for all we knew, deadly. Taking our cues from baboon and bird leftovers, we soon learned the astringent, pungent, and stomach-churning plants to avoid. After weeks of trial and error, we identified and named seventeen edible species within an hour's walk of school.

Shirley found the same caution in the Pumphouse Gang she translocated from the Rift Valley to a new home in Laikipia Plateau north of Mount Kenya in 1984. At first the Pumphouse baboons mostly stuck to familiar plants, by cautious trial and error adding new species picked up from local males transferring into the troop. Like our bush gang, the young males were the most exploratory. I felt a touch of déjà vu watching an adolescent test out an unfamiliar root tuber dropped by an adult male. After a few tentative sniffs and nibbles, the youngster hurriedly dug up fresh tubers to get a jump on a juvenile following close behind. In another incident Shirley found baboon youngsters *gekking* in alarm at a strange object in the grass: a worn-out tennis shoe.

Humans outcompeted other primates in Africa and eventually all mammals by testing out new foods and expanding their diets. The sheer diversity of foods our ancestors ate, from insects to whales and from algae and fungi to plants, leaves, fruits, nuts, berries, and roots, carried risks. How could our forebears possibly have learned and remembered all the edible and poisonous foods? The elaborate ritual Tukanoan farmers of the Amazon use to detoxify poisonous manioc roots points to the gradual accrual of know-how over generations handed down through cultural practices and rituals.

Our ancestors also borrowed and harnessed the toxins of other species of animals and plants to kill prey and cure illnesses. The Wata hunters of Kenya were highly proficient at rendering down a tarlike concoction from the roots of the *Acocanthera* tree and testing its potency by nicking a vein to see how fast the poison turned the trickle of blood black before scraping it off. Plastered around the arrowhead and wrapped in leaves, a single shaft shot into the gut of an elephant can bring it down in a few hours.

When I took on David Maitumo as my field assistant, I asked him to tell me the Maasai names of plants and teach me their uses. His list ran to over two hundred species, including fever tree sap, used as chewing gum; *Maerua* berries, to supplement the Maasai staple milk and meat diet; *Salvadora* twigs, used as toothbrushes; and *Acacia nilotica* (*ilkiloriti*) roots, used to speed the digestion of warriors gorging at an *orpul*—a meat feast. The coarse grass stems of *Sporobolus consimilis* are used as a screen in the family home, and *Acacia mellifera,* the wait-a-bit thorn, for fencing in livestock and keeping out predators. The list included medical concoctions for every imaginable ailment from morning sickness to headaches and diarrhea.

Our ancestors were astute pharmacologists who knew which plants and animals cured fevers, killed pests, caused hallucinations, or served as a pick-me-up. Medicine men and shamans who garnered an encyclopedic knowledge of the curative properties of plants systematized folklore cures in many cultures, giving them a revered and feared place in society. The skill and reputation of trustworthy healers were often learned and earned by renowned families and clans specializing in herbal medicine.

We owe much of our modern pharmaceutical products to traditional medicines. The most widely used drug, aspirin, was extracted from the willow and salicylate-rich plants as early as the second millennium BCE in Egypt. Taxol, a cancer drug, was originally discovered in the Pacific yew of North America, and the malarial drug quinine came from the cinchona tree in the Amazon. About a quarter of all modern pharmaceutical products are derived from plants, bacteria, or fungi.

The interplay of our inborn caution and our urge to expand our diet and medicine chest is rooted in our nature and culture. Despite the claims of paleo diet advocates and homeopathic practitioners, natural products can be just as dangerous to our health as processed foods and manufactured drugs. We've always had snake-oil salesmen and tricksters promoting elixirs and the fountain of youth. The only way to distinguish safe from hazardous foods and drugs is by testing them. How can we do so in our global age of tens of thousands of natural and manufactured foods flooding the marketplace around the world?

Just as we have a natural revulsion to snakes and spiders, we have an inbuilt aversion of our own bodily emissions—shit, piss, farts, and snot. We hate the stench of feces and avoid crapping and pissing on the floors of our homes and farting in company. Our dirty-nest sensitivities help rid us of parasites, disease, and pests. Our sensitivity to environmental pollution evolved in lockstep with village and urban life. We are fussier in permanent settlements and confined spaces than in mobile camps and around the campfire. African villagers scrupulously sweep their compounds to rid them of feces and trash and to avoid rodents, snakes, pests, and bugs. In Swahili *unadthifu*—cleanliness—is a virtue, *uchafu*—dirtiness—a vice.

The first written code to protect the public health from waste, pollution, and disease in cities dates to the Babylonian Code of Hammurabi in 1770 BCE. Cleanliness is next to godliness in European folklore, and with good reason. The densely packed towns and cities of the industrial age were as fertile breeding grounds for diseases as were medieval towns and cities.

Water-borne diseases like cholera, vector-borne diseases like bubonic plague, and contact diseases like anthrax flourished in sprawling towns and cities with poor waste disposal and sanitation. The freedom of the urban commons becomes a tragedy for all: a density-dependent killer.

In the global age, our self-inflicted pollutants have grown exponentially with the mountains of garbage, chemicals, toxins, heavy metals, and waste gases we produce. Our senses are not designed to detect these inert compounds. As processed foods replace our natural diet, the list of potentially harmful ingredients has grown with the battery of artificial dyes, preservatives, and additives of all sorts. Public-safety bodies like the Environmental Protection Agency have replaced the trial-and-error methods of traditional medicine men with scientific methods of detecting our modern harmful products, such as the insecticide DDT and BPAs in the plastics that permeate our environment.

If the regulatory, monitoring, and enforcement agencies responsible for public safety were scrapped, deaths and ill health from toxins in our air, water, soils, and food would soar. A laxity in food standards and monitoring in China in the early 2000s resulted in acute sickness in three hundred thousand infants, fifty-four thousand hospitalizations, and several deaths from melamine added by suppliers to make milk appear more nutritious.

Whereas once we worried about visible city smog choking our lungs, now far smaller ultra-particles are proving lethal. Measuring less than two-hundredth the width of a strand of hair, ultra-particles are undetectable by conventional monitoring tools. Such tiny particles, emitted by burning fossil fuels, wood, and cigarettes, invade the nasal cavity and brain, where they damage neural cells and may account for a fifth of dementia cases worldwide.[1]

Thousands of industrial by-products dumped into the global commons are compromising the natural capacity of the atmosphere, oceans, lakes, rivers, and soils to degrade and recycle pollutants. Nitrates and phosphates leaching off farms into the Mississippi watershed and into the Gulf of Mexico have caused a vast oxygen dead zone, costing the local fishing and tourism industry an estimated $82 million a year. Halfway around the world in Japan, the Ministry of Trade and Industry has put the cleanup cost of the 2011 Fukushima radiation leakage at $180 billion.

The waste products of our global age are so pervasive and our monitoring skills so paltry that several threats to planetary health have been discovered by sheer luck and the doggedness of scientists like Charles Keeling. A geochemist at the Scripps Institution of Oceanography in San Diego, Keeling set up a monitoring station on the peak of Mauna Loa on Hawaii's Big Island in 1958 to determine baseline atmospheric CO_2 levels free of urban smog. Continued unbroken ever since, the Mauna Loa station recorded CO_2 levels ratcheting up from less than 320 initially to over 400 parts per million in 2016. How much higher would CO_2 levels have risen before the Kyoto Protocol struck a global consensus to cut greenhouse gas emissions had Keeling not set up his monitoring station? How much wider would the atmospheric ozone hole have spread and the incidents of skin cancer risen had Sherwood Rowland, Mario Molina, and Paul Crutzen not demonstrated the impact of chlorofluorocarbons on ozone destruction, or had British Antarctic Survey scientist Joseph Farman and colleagues not discovered the gaping hole in the ozone over the Southern Hemisphere, prompting the Montreal Protocol to ban the causative refrigerator coolants?

Only a few of the thousands of chemicals we encounter in our daily lives have been thoroughly tested for their harm to humans and the environment, and fewer still for their long-term impacts and cumulative effects through the food chain.

The list of dirty nest pollutants grows longer and the impacts more pernicious with worldwide economic growth. The antibiotics doctors overprescribe for patients and the agroindustry injects into cattle to stem infections picked up in the quagmire of feedlot manure create an arms race in which ever-stronger antibiotics are needed to combat the accelerating resistance of bacterial strains such as methicillin-resistant *Streptococcus aureus*. We are creating a young generation with lower immune capacity, greater dependence on drugs, and higher incidences of asthma and allergies. The exposure runs the risk of disrupting our hormone system. More troubling, scientists are discovering that the balance of bacteria in our guts—the microbiome that regulates our digestive system—can affect our moods and health when out of kilter.

Plastic waste is the latest and visibly most alarming pollutant sullying our planet. Over 380 million tonnes, half from single-use products, are manufactured each year. Plastic bags, bottles, plates, straws, tables, chairs, packing materials, and hundreds of other products litter lands, rivers, and lakes around the world. Thousands of livestock each year in Kenya alone die from guts compacted with plastic bags. Floating islands of plastic trash float down our rivers. Discarded plastic flutters across the landscape, is trapped like tumbleweed by fences, buildings, and crops, and festoons trees and bushes in our remotest parks and recreation areas.

Over 8 million tonnes of plastic enter the oceans each year, creating giant gyres of floating plastic in the central Pacific. Dozens of whales are washing up on beaches from the Mediterranean to the Philippines, their stomachs filled with thirty to sixty kilos of plastic bags, cups, and shopping bags. Plastics break down into micro debris, which forms a thin film over the ocean floor. The debris is sucked in by filter-feeding bottom inverte-brates and works its way up the food chain into fish and piscivorous humans. Plastics may take hundreds of years to decompose into dozens of toxic substances, including diethylhexyl phthalate, lead, cadmium, and mercury. Imagine the cost to our children, even long after we may have found safer substitutes.

How can we be sufficiently alerted to the scale of our global dirty nest soon enough to act?

As a kid I was dazzled by the photos of the desert blooms in the *Arizona Highways* magazine. On my first trip to the United States in the early 1970s I rented a car and drove down to Organ Pipe Cactus National Monument to hike the desert trails. I was shocked at the trash and beer bottles littering the highway, the warning signs of deer crossings riddled with bullet holes, and the iconic saguaro cacti of cowboy movies blasted with buckshot. The land of the brave had yet to shuck its endless frontier and Wild West mentality.

Back to meet the Malpai Ranchers twenty years later, I was astonished by the clean highways and the evident pride of Arizonans. Drivers brash enough to dump trash or shoot up a deer sign faced hefty fines and honk-ing cars. *Adopt a Highway* signs posted along the freeways extolled the social responsibility of corporations keeping Arizona's highways clean. The state's culture had changed from crass abuse of nature to pride and caring in a few years. Turning around our frontier mentality when it comes to our shrinking planet will take an Arizona change of culture on a global scale.

By the turn of the millennium, the U.S. Environmental Protection Agency (EPA) was moving away from centralized government control of pollution to an array of new tools befitting the global age, according to a 2002 report by the National Research Council (NRC) of Washington.[2]

The tools included education, incentives, reputation, and voluntary action. What I find most intriguing about the NRC report is a short statement noting that the new national EPA tools hark back to successful community responses. The nature of our emissions has changed, the report goes on, from a few pollutants highly concentrated in the smokestack industries of the nineteenth century to the tens of thousands of pollutants spread globally and among billions of households by the new industries of the twenty-first. The new pollutants found in plastic bottles, for example, include bisphenol A (BPA), which can cause brain and behavioral changes in infants and high blood pressure in adults.

In recent decades a new field, industrial ecology, has sprung up. Industrial ecology looks at the urban and industrial landscape as a human ecosystem. A good example is determining injurious levels of toxins and appropriate emission standards. Paracelsus, sixteenth-century physician, said all things are poisonous; it is the dose alone that makes them safe. Using the Paracelsus creed, toxicologists have assumed the response of animals to a toxin is proportional to its dose. This was a fair assumption for most pollutants and was used to set safe standards of exposure based on toxic levels in lab rats. The linear dose-response curve doesn't hold, however, for a battery of new pollutants. The estrogen-mimicking p-nonylphenol, which disrupts female hormone cycles, for example, is most damaging at low and high doses; BPA is most dangerous at moderate doses.

In Rachel Carson's day, the chemicals dangerous to humans and animals were far fewer. They included DDT, mercury, heavy metals, and synthetic drugs like the sedative thalidomide, which if taken by pregnant women could cause grotesquely shrunken limbs and deformed bodies in newborns. Today the number of industrial chemicals is eighty thousand and rising. Exposure to deleterious secondary effects from the interactions among the battery of chemicals adds another layer of complexity. The ripple effects can be erratic and hard to predict; the causes of ill health can be difficult to pin down, even in the case of well-studied diseases such as type 2 diabetes.[3]

How can we possibly measure the dose-response curves of tens of thousands of chemicals and their interactions in lab animals, let alone humans, who live forty times longer and accumulate many more toxins, carcinogens, and hormone disruptors in their organs during a lifetime? The hormone-disrupting pollutants rampant in our food, electronics, furniture, and personal care products amount to $557 billion in disease costs in the United States and Europe.[4]

Epigenetic effects—gene expressions modified by environmental factors—can be transmitted from parent to offspring over several generations. The stunted growth of famine victims and the obesity caused by rich diets, for example, can affect the health of children for a couple of generations. The knock-on effects are hard to detect and harder to prevent when inherited by our descendants.

Despite living longer than ever, we are increasingly likely to die of Western diseases, as British doctor Denis Burkitt called them when working in West Africa in World War II, rather than of starvation or contagious diseases such as malaria, TB, and typhoid.[5] Medical bills consume a growing portion of the national budget and old-age savings in the industrial world. The deferred costs of the industrial age live now, pay later generations have arrived: we are paying the price and handing down a far larger burden to our children.

Cleaning up our twenty-first-century dirty nest calls for big funding, international collaboration, and the analyses of huge data sets made possible by the networked open source capacity of the information age. The European Commission, for example, has launched a study involving thousands of voluntary subjects carrying smartphones equipped with sensors to measure hundreds of chemical agents. Scientists analyze blood samples for molecular changes and biomarkers of damaging exposure. Another study is testing the effects of chemical agents on the growth rates, obesity, asthma, and immune systems of pregnant women and children.

The sooner we detect our self-inflicted wounds, the earlier we can act. Recycling trash as gas, breakthroughs in solar, wind, and wave energy, catalytic converters to reduce exhaust emissions, bacteria to digest oil spills, and offset schemes to "cap and trade" carbon—all help in curbing emissions. The changing nature of production can be tackled with better technology and management, according to the NRC report. Between 1955 and 1980 the United States added 40 million jobs, largely in service industries, health care, communications, wholesale, finance, insurance, and real estate—only 10 percent were in manufacturing. Better product design, the use of lighter materials, recycling, and a shift to IT technologies able to boost production efficiency and cut waste and emissions must be part of the economic and behavioral shift away from a carbon-based economy.

No less important is the need for new monitoring and auditing methods to track and curb the national carbon footprint accounted for in offshore manufacturing.

Despite the falling quantity of materials used in producing manufactured goods, consumption is still rising worldwide as more people board the global express in response to the Lebo Effect, described in chapter 13. The developing nations will need to leapfrog to the fourth industrial revolution to prevent an ecologically disastrous world-sized America. It is in the interests of the catch-up nations to do so. China surpassed America's annual carbon emissions in 2007 and currently vents twice the amount into the atmosphere. If China used its rich coal reserves to follow the U.S. path to industrialization, its CO_2 emissions would be four times America's at its peak emissions in the late 1990s. As it is, China is converting to clean energy by investing $320 billion in renewable sources, aiming to achieve its emissions reduction under the Paris Agreement by 2035. Adopting the new clean energy route to industrialization will create 13 million new jobs and combat the deadly smog choking Beijing and northern China. In achieving its goals, China will become the world leader in exports of solar and wind energy technology and, in the process, reverse its offshore carbon footprint by reducing emissions in countries buying its clean energy technology.

Pollution controls must be tackled locally and coordinated nationally and globally to curb global emissions. Disposing of the trash and toxic waste of rich countries in poor nations as an economically attractive transaction, as former chief economist of the World Bank and president of Harvard Larry Summers once proposed, not only ignores risks to the poor, who comb dump sites for saleable scraps, but sidesteps domestic environmental standards and does nothing to reduce the global footprint. The "not in my backyard" (NIMBY) view of waste disposal prevalent in the 1980s fails to create the skills and responsibility for solving twenty-first-century pollution.

Domesticating pollution control to zero the emissions dumped into the global commons focuses national remedies on the circular economy. Pollution must be incorporated in the production system, not dumped out the back door as waste, trash, and toxins. Some solutions are delightfully simple, imaginative, and artistic. Flip-flop sandals littering the beaches of Kenya, for example, are collected, laminated, and carved by artisans into decorative lions, elephants, planes, and boats and sold at craft fairs. Plastic bags and bottles are collected by thousands of unemployed workers and sold to recycling plants for 45 cents a kilo.

Corporate and individual responsibility are key to the mix of pollution abatement. At a national level, the U.S. Freedom of Information Act of 1967 and the Emergency Planning and Community Right-to-Know Act of 1986 have reduced toxic chemical emissions by mandating the release of corporate information on waste and toxic disposal under the Toxics Release Inventory regulation. The information has increased the detection of deleterious emissions by corporations and, in the process, their responsiveness to public pressure.

The impact of our emissions on wildlife can be used as an early warning of pollution threats to our own health. In 1956, for example, a "minamata" plague caused neurological damage and deaths in thousands of Japanese in the village of Minamata. The culprit was traced to a mercury by-product dumped into the bay by a chemical plant owned by the Chisso Corporation. The mercury concentrated through the food chain, including in dolphin and pilot whale meat eaten by the local populace. Minamata became a symbol of corporate environmental abuses and made a powerful case for banning the dolphin slaughter. In the United States the near-extinction of the bald eagle from eggshell thinning caused by DDT alerted Americans to the dangers of toxic insecticides and precipitated mandatory testing, regulation, and monitoring of insecticides and pesticides.

The "canary in the coal mine" used to detect and warn miners of lethal levels of methane deep underground became a metaphor of our impact on the environment and the environment's effect on our own health during the environmental movement of the 1970s. Monitoring the toxin levels and health of wildlife indicator "canary" species is a more effective and humane way of tracing the cascading effects of pollutants through an ecosystem than testing the dose-response of laboratory rats and primates. The health of wildlife everywhere—polar bears in the Arctic and tuna in the Pacific—is a powerful metaphor for a healthy planet.

The canary metaphor takes us a step closer to linking the welfare of humans to the health of indicator species but falls short of highlighting the importance of biodiversity in maintaining a functioning biosphere. This calls for another step linking human well-being to planetary health and the value of the services nature provides us.

19

Nature and Human Well-being

Watching baboons with Shirley is like looking through a rearview mirror at our ancestors' first venture onto the savannas. There is no mistaking the males' swagger and strut as they vie for rank. Shirley had named a baboon Jonah, after me, and naturally I follow his progress avidly as he strides through the troop with all the confidence of a sergeant major. Females are far subtler than males, reflecting their inherited rank and baboons' stable dominance hierarchy. Peggy, the top-ranking female, was born into the top-ranking family. She is calm, gets her way using rank and social savvy, and tips the balance in favor of her own offspring in family rivalries.

We humans flaunt our stuff too, of course. Generals show their brass, presidents drive around in motorcades, and warring New Guinea Highlanders brandish their meter-long wooden penis shafts as large as an elephant's—though far more decorous, I would add. For the most part, like Peggy, we express our rank in subtler ways, by embellishing our natural attributes using haughty accents, coiffured hair, rouged lips, crimped waistlines, pumped-up boobs, and steroid muscles.

Charles Darwin, in *The Descent of Man*, developed a theory of sexual selection to explain the evolution of the flamboyant color, song, and dance in animals—the peacock's majestic feathers, the sweet melody of canaries, and the elegant dance of whooping cranes.[1] Most electrifying of all is the dance of the satin bowerbird outside its courting chamber made of feathers, shells, bones, grasses, and leaves—and nowadays brightly colored shards of glass and plastic.

Darwin's notion that the elaborate displays and rituals of animals boosted their reproductive success through mate choice has stood the test of time. Israeli scientist Amotz Zahavi went a step further: overt displays such as the peacock's hefty feathers impose a handicap on males and thus signal their fitness.[2] David Rothenberg suggests that song, dance, and ornamentation in animals can be selected as fads and styles—independent of their reproductive value—by stimulating pleasure centers in the brain and attracting mates.[3] Cornell ornithologist Richard Prum, who has spent decades studying the ornate colors and courtship of birds in the tropics, comes to a similar conclusion: beauty and elegance evolved for their own sake because of their powerful attraction to females.[4] The case is far stronger, of course, in humans. Artists, poets, writers, and scientists today are admired more than bow hunters.

When it comes to humans, evolutionary biologists have a hard time persuading us that our love of art, music, and dance is merely reproductive bait. The entertainment and fashion industries are booming as birth rates plunge. Freed of food gathering and breeding for the most part, we indulge our senses, savoring beauty, scents, and sounds that trigger our endorphins and pleasure centers. True, showing off our prowess in sport has its rewards in status and mates, but for most of us, playing games is self-satisfying and pleasurable to the point of Nintendo addiction.

We enjoy museums out of a fascination with the natural world and our cultural heritage. Even the most basic utensils like knives and forks have become works of art. In the Byzantium section of the New York Metropolitan Museum, the Baroque writing bureaus inlaid with ornate mother-of-pearl, gold ribbings, and delicate curlicues became objects of beauty and symbols of wealth. At the San Giorgio workshop in Venice I made a point of visiting the famous spandrels of the San Marco Cathedral, not so much for their artistry as for the story they told of biologists overinterpreting the tiniest animal and plant variations as finely honed and adapted products of natural selection.

Paleontologist Stephen Jay Gould and geneticist Richard Lewontin took exception to the adaptationist paradigm, as they called it—the view that every animal trait has a purpose and is a perfected outcome of evolution.[5] To illustrate the point, Gould and Lewontin noted that the ornate ribbings and delicate overlaid artwork on the load-bearing spandrels of the San

Marco church in Venice served no structural purpose. Like the inlays and curlicues of Baroque writing bureaus, the artists simply took advantage of the free surface on the spandrels. The same is true of evolution. Not all traits are selected by evolution or have a function. Dr. Pangloss claims in Voltaire's *Candide* that our protruding nose is nature's design for wearing glasses and our shoulders for wearing braces to hitch up our pants. More prosaically, our nose resulted incidentally from selection for a shorter jaw as our ancestors discovered how to pound and cook fibrous plants to make them more digestible, and our shoulders from standing upright.

Like the spandrels of San Marco, our emotions and senses inherited from Miocene apes and refined in Pleistocene hominids have become exaptations—traits co-opted for uses other than those originally selected. We have co-opted all manner of traits for new uses. Freed from everyday hunting and food gathering, we redeploy our big curious brains, pleasure centers, and innovative skills in learning for the joy of it, reading literature and indulging our fancy in art, music, gardening, and of course sex, for fun rather than reproduction.

We are emerging from a bounded ecology and bonded communities onto a new global stage and into our self-made Human Age. We are rearranging our environments and embellishing them with images, sounds, and smells to stimulate our senses, emotions, physicality, curiosity, and creativity. The old ways are alien to all but a few subsistence societies still living on the natural sustenance of the land.

For over half the world the city has become the new landscape and ecosystem. When I arrived in England fresh from Tanzania in 1960, I entered an alien habitat on my first visit to a department store, Bentalls, in Kingston upon Thames. Overwhelmed by the jostling crowds, the abundance of jewelry, perfumes, clothes, appliances, and foodstuffs, and the welter of unfamiliar sights, sounds, and smells assailing my senses, I was too flustered to find my way to the winter-wear department.

I see the same bewilderment and confusion in first-time visitors on safari. Their senses are so saturated with unfamiliar sounds and sights they imagine every stump is an animal and lose themselves in the bush, where every tree looks alike. Bill Calder, a renowned hummingbird physiologist, took me on a hike into the mountains outside Tucson, Arizona, when I visited him in the 1980s. There are bighorn sheep here, he told me, but they blend in so well I've never seen one in thirteen years. Within half an hour I picked out a small herd high on a rocky promontory. Even with my directions Bill, who could spot a hummingbird with hawklike acuity, had trouble picking out the sheep among the boulders.

For many of us, the city has become our native habitat and ecosystem: nature is remote and alien, worth visiting as a curiosity. The travel and tourism industries generate over $7 trillion a year—9.5 percent of the global GDP, are growing at 4.5 percent a year (twice the world's population growth rate), and support 266 million workers.[6] Travel dwarfs the $1.93 trillion entertainment industry, the $146.5 billion sports industry, the $51 billion art industry, the $15 billion music industry, and the $80 billion literature industry.

Tens of millions of people around the world watch wildlife and nature documentaries on TV and up-to-the-moment happenings on YouTube. Outdoor recreation reinforces rather than competes with economic growth; according to the Outdoor Industry Association, outdoor recreation generated $887 billion in consumer spending in 2017, 7.6 million jobs, and $65.3 in state and federal taxes.[7] Nature and outdoor tourism feature far larger in many developing economies. Kenya's tourism industry, 80 percent of it attributable to wildlife, employs 11 percent of the country's labor force and generates 12 percent of the nation's GDP.

On the downside, the outdoor and nature industry benefits the wealthy and not the poor. The wealthy travel at their leisure and enjoy nature for pleasure. The poor clear natural habitats for farms, cut down trees for firewood, and chase off crop-raiding monkeys. The biodiversity-poor rich nations look to the biodiversity-rich poor nations to experience the nature they have destroyed, and the rural poor they pass along the way are striving to escape nature and enjoy the lifestyle world travelers take for granted. These two worlds, the locomotive and caboose carriages of the global express, have different outlooks, values, and conservation priorities. The rich world is expanding the meaning of conservation to include biodiversity and cultural heritage as the poor world is struggling to escape them. The challenge of globalization is to find a place and space for wildlife in ways that satisfy both ends of the global express.

The notion of biodiversity is a product of modern science, knowledge, and sensibilities. Kerenkol couldn't understand my love of wildlife. For him the value of wildlife was primarily as second cattle. He was too bound by nature and tied to his livestock to share my freedoms in life, yet I suspect he got greater fulfillment from his cattle than I do from wildlife. Kerenkol expressed his joy by singing to his cows, decorating their bodies, and petting them. Out in the bush I often sing too, out of the sheer joy of being alone with wildlife, but I don't share Kerenkol's deep intimacy with animals. If I did, wild animals would lose their wildness and the values I so cherish.

I readily admit that Kerenkol's life was far richer than mine. I often watched in awe as his entire world converged on his family, friends, and livestock in his corral each evening, giving him a sense of worth and recognition modern living lacks. Kerenkol's way of life has done more to conserve wildlife than conservationists who admonish the Maasai for being wedded to livestock. The claim of British biologist and writer Julian Huxley in the 1960s that a country without parks can't be considered civilized speaks to the failure of his own nation to coexist with wildlife—and a need to atone for Britain's rapacious past by pressing for wildlife reserves in Africa. I should add that Huxley's Britain at the time had yet to set aside any of its own land exclusively for wildlife.

Britain, whose expansive forests once harbored wolves, bears, and bison, has become an island of farms dotted with cattle and sheep, a maze of roads, and cluttered cities. For all its domesticity, the bucolic rural English countryside was the inspiration of the romantic poets Alfred Lord Tennyson and William Wordsworth, the sublime artists Thomas Gainsborough and John Constable, and the infatuated naturalists Gilbert White and Charles Darwin. Darwin claimed on his return from the voyage of the HMS *Beagle* that Britain probably exceeded in beauty anything he beheld around the world. Britons, once noted for their cruelty to animals, have become a nation of animal lovers.[8] Natural England and the National Trust have conserved a unique blend of natural and cultural heritage in the rural countryside. Even Britons, though, have their limits on how close and wild they want their nature. Reviving the otter and reintroducing the beaver are easy steps. Reintroducing the wolf, bear, and bison is a step too far, at least for now.

Britain's cultural transformation from abject cruelty to animals in the pre-industrial age to a nation of birders who build underpasses to protect frogs from highway traffic shows that a love of nature is not inborn. Biophilia doesn't ring true to nations still wrestling with the raw forces of nature, subdued by the West. Biophilia is a spandrel of San Marco, a fanciful adaptationist's hope that we can reverse our plundering past with an ineffable love of nature.

The contrast between herders and farmers struggling to live with wildlife and tourists enjoying a game drive collides starkly in Amboseli. Cynthia Moss's portrayal of Echo the elephant's family life as humanlike captured the hearts of animal lovers around the world.[9] The economic benefits of wild animals and the protection the Maasai have given them make Amboseli's elephants the safest in Africa and world renowned. Against the continental backdrop of twenty-five thousand elephants a year being slaughtered for their tusks since the international one-off ivory sale in 2008, Amboseli's herd moves placidly across the landscape, protected by community scouts funded by conservation organizations like Big Life Foundation, International Fund for Animal Welfare, Amboseli Trust for Elephants, and African Conservation Centre, backed by Kenya Wildlife Service rangers.

For all the evident success of community-based conservation in Amboseli, elephants risk losing the local support they've gained. In 2014 the crop losses to the local community exceeded $500,000. Daniel Leturesh, chairman of the Amboseli Ecosystem Trust, the person who talked down warriors intent on spearing a lion in reprisal for an attack on an age mate, led a crowd of eight hundred Maasai in celebrating the launch of the Noonkotiak Community Resource Centre bordering the park. This is our center for coordinating the work of our Lale'enok scouts and overseeing the Amboseli Ecosystem Management Plan, he said. The Maasai have been the real custodians of wildlife and yet suffer most. Lions and elephants have grown in numbers and no longer fear people. Children can't go to school, farmers and herders are losing crops and livestock, yet where is the compensation long promised by government? Our tolerance of wildlife is fading for elephants and lions.

We can't rely on an unproven affinity for nature or expect those who suffer from wildlife to share the warm fuzzy feelings of people like me, immune to nature's hazards. Our modern romanticism of nature emerges from our ecological emancipation, new sensibilities, and leisure time. Urban dwellers, free of wildlife hazards and with money to spend and time to spare, are drawn to nature for its recreational, educational, adventurous, romantic, and spiritual values. Over 90 percent of Kenyans living in cities see value in wildlife. Among rural communities suffering depredations and getting nothing out of wildlife, 90 percent want to be rid of it and take over parks for herding and farming.

Our appreciation of nature grows with education, urbanization, and our newfound admiration for places untrammeled by humanity. The efforts of the Wildlife Clubs of Kenya over the last forty years have changed attitudes toward wildlife among Kenyan youths from antipathy and indifference to an advocacy movement. Thousands of kids marched through the streets of Nairobi in support of an ivory ban in the 1980s. The national Parks for Kenyans campaign I launched at the Kenya Wildlife Service in 1997, coupled with lowered entry fees for citizens, saw a surge in Kenyan adult visitors to see wildlife close up in parks, rather than remotely on TV.

The fast pace of life in the cities of Europe and North America makes the Maasai way of life—living off their herds in the natural world—seem far simpler and happier to the urbanite. The eulogizing of nature, like the appeal of Stone Age diets, ignores the Lebo Effect—the lure of better prospects and richer lives. Paradoxically, the stresses of modern living eventually draw the new urban recruits back to the wild places urban living has created.

This, at least, is the message Richard Louv, a renowned advocate of the curative powers of nature, is promoting. I was keen to talk to him about his ideas when in 2015 Shirley arranged for me to meet him and Paul Dayton, a well-known marine ecologist and conservationist, for lunch on campus at the University of California, San Diego. Richard has written two best sellers—*Last Child in the Woods* and *The Nature Principle*.[10] *Last Child* has sold over a million copies and attracted a devout league of followers to Louv's campaign to get kids into nature. When we met, he wanted to hear about my childhood encounter with a sable antelope and how the experience turned me from hunter to conservationist.

Like E. O. Wilson, Louv believes a biophilia instinct draws us back to nature because when we are deprived of it, we suffer a nature deficit disorder. Over the course of a long and riveting talk about biophilia, I was amused at our role reversal: he, the writer, thinks biophilia is a genetic propensity; I, the biologist, think love of nature is a product of upbringing and culture.

Despite our different takes, Louv and I find common ground in nature stories and the reticence of biologists to admit the love of nature that drew them to field research—as if a show of emotion compromises their scientific integrity. It probably does in scientific meetings and publications, but nothing thrills kids and the general public quite so much as scientists recounting enchanting stories of their discoveries and escapades in the field saving threatened species. Richard has hit a golden vein in collecting tales of scientists' encounters with animals, and I feel comfortable telling him about the passions, adventures, and curiosity that made me the conservationist and scientist I am.

Louv's back-to-nature elixir for nature deficit disorder appeals to digital-age kids suffering from depression, obesity, and sedentary ailments. He has assembled a vast body of literature extolling the benefits of the outdoors on physical and mental health, inspiration, education, spirituality, and well-being. Louv's outdoor advocacy for nature-deprived kids works just as well in Africa as America—perhaps better because of the abundance of wildlife and a legacy of human-animal coexistence. Schoolkids in East Africa thrive on safari and love to hear their parents' folktales about the greedy hyena, clever hare, and gentle elephant. Nobel Prize winner Wangari Maathai's Green Belt Movement has made a generation of Kenya women and children ardent tree planters and inspired millions of people around the world.

The joy that comes from contact with nature and other creatures cannot be matched by building a downtown ballpark, community hall, temple, museum, zoo, or meditation center. These are human constructions devoid of Louv's nature values. But we can't expect to set aside half the planet as protected areas, as E. O. Wilson advocates in *Half Earth*.[11] The nature constituency is far too small, the loss of farm and ranch lands too great, and the displacement of people too disruptive and unethical for that anytime soon. Most people, even ardent nature lovers like Henry David Thoreau, spend little time in the woods anyway, because they have other lives to lead. In the industrialized world, nature represents not a way of life but recreation, freedom, curiosity, romance, spiritual retreat, or adventure. Whatever the appeal, the call to nature is like the call to prayer, a sanctuary from daily life.

Modern environmental and nature sensibilities took two hundred years to emerge in the West.[12] The languid pace is no reprieve for the rest of the world. Closing the gap depends on relief from hunger and poverty, gains in productivity, thriftier use of resources, improved environmental health and security, and investing in youth education and citizen access to nature. On the bright side, the large proportion of young people in developing countries, and the speed at which they are becoming the dominant voice and culture, will hasten general acceptance of Louv's nature values such that it will come to pass in a matter of decades rather than centuries.

If the public demand for nature as a tonic for improving human well-being offers the best hope for protecting open spaces and wildlife in the long run, how can we make biodiversity a universal and household value in the short term?

20

Natural Capital

Elephants are the largest land herbivore and, as befits their size, play an outsized role in shaping habitats. On a trip to Central Africa in 1986 to assess the status of the forest elephant, I was riveted by the part they play in diversifying forests by dispersing seeds, opening and perpetuating glades, and creating patchiness in the age structure and composition of plants. In a *Discover* magazine article covering the trip, I called elephants the architects of the forest.[1] Elephants wandering freely create a patchwork forest rich in plants, mammals, and birds. Where elephants die out, the forest falls silent. The analogy, as with any analogy, is only part of the story. Geology, topography, climatic swings, fires, hurricanes, and above all humans also play orchestrating roles.

After that forest trip, I saw the role of elephants shaping the ecology of Amboseli with fresh eyes. My early studies had shown elephants turning Amboseli's woodlands to grasslands when compressed into the park, confirming other studies of harried herds across Africa from the Murchison Falls in Uganda to Kruger National Park in South Africa. The bulldozer-like impact of elephants in African parks is an artifact of human disruption rather than the animals' natural behavior. Once elephants fleeing poachers and settlers crowd into the safety of parks, the ecological minuet of their natural movements becomes a repetitive stomp. In evading human predators and trapped in parks, elephants become like the feral goats on the Channel Islands of California, destroying the vegetation and impoverishing the landscape.

In the late 1980s I was utterly flummoxed flying into Mikumi in Southern Tanzania, a deciduous woodland where we had hunted elephants thirty years earlier. I checked my map to be sure I hadn't botched my navigation. Scouting around, I was reassured: the familiar Uluguru Mountains loomed to the south, the Mkata River to the north. But where were the miombo woodlands we had slogged through on our foot safaris, swatting off tsetse flies? Thirty years on, the woodlands of Mikumi had become a mini Serengeti dotted with wildebeest and zebra.

I had seen the reverse ecological changes in the Selengei north of Amboseli. Here thinly scattered bushes had smothered the grasslands when poachers gunned down the elephant population in the 1970s. Maasai settlements that once allowed herders to reach the distant grasslands had become fading scars in a thicket of drab gray bush.

Primed by the Central African forest trip, I delved into the role of elephants in shaping the savannas. Surely, given their huge ecological impact, elephants should have an outsized imprint on animal communities too? By good fortune I had a ready-made natural experiment to test the assumption—a gradient of elephant densities stretching from the compressed herds in Amboseli National Park to the poaching-depleted herds to the north.

Sure enough, the ecological footprint of elephants was unmistakable. In the poached-out areas to the north, a few species of acacia had crowded out the grasslands. In the center of Amboseli National Park the compressed herds had turned woodlands to grasslands, and the once-diverse inventory of plant species was overshadowed by a resilient few. The richest assembly of habitats and species coincided with the interface of elephants and livestock on the border of the park. Here, where elephants crossed out of the park to browse the bushlands at night and cattle filed into the park to graze the grasslands by day, the ecological minuet played out in miniature, creating a patchwork of woodlands, scrublands, and grasslands rich in species.[2]

The keystone role of elephants in shaping habitats and the wealth of species in Amboseli showed up in bird counts made by primate researcher Jeff Walters in the 1970s and was repeated in counts undertaken for me by ornithologists Duan Biggs and Dana Morris thirty years later. The greatest wealth of bird species occurred in the patchwork of habitats created by the interplay of elephant and livestock. I often returned to the site where I first pitched my tent in a stand of yellow fever trees in 1967 to get an acoustical feel for the changes. I would sit in my camp chair enjoying the constant hooting, screeching, honking, and chattering of owls, eagles, falcons, hornbills, woodpeckers, hoopoes, starlings, and dozens of other bird species. Fifty years on, the short-grass plains created by elephants destroying the woodlands are hauntingly silent, except for the murmuring wind, the twitter of seed-eating passerines, and the occasional harsh *yak-yak* of crowned plovers.

Queen Elizabeth is no ecologist, but she was shocked at the loss of forest at Treetops Lodge in Abedare National Park in the Kenya Highlands on returning forty years after receiving news of her father's death and ascending to the British crown while staying overnight at the lodge in 1952. John Waithaka, a former student of mine and now chairman of the Kenya Wildlife Service, researched the gargantuan changes and concluded that elephants, blocked from migrating to the lowlands by an electric fence and lured as a tourist attraction by the salt lick at Treetops, had turned the forest to grassland.[3]

Plowing through the sightings of animals itemized in the daily visitor log at the lodge, John found an even more astonishing subtext. The earliest entries recorded an abundance of primates, squirrels, and other tree-dwelling mammals. As the forest thinned and the understory thickened to scrubland, giant forest hog and bongo sightings rose before giving way to zebra, eland, and plains animals as the grasslands spread. The diversity of species rose with the decline in forest cover, peaked during a mix of forest, scrubland, and grasslands, and declined as the plains spread. Abedare, an island park, is far too small for a natural ecological minuet of animal movements creating habitat diversity to play out.

The Amboseli and Abedare story is a replay of the biodiversity loss across Africa, wherever poached and harassed elephant herds abandon their migrations for the safety of protected areas. The loss of biodiversity hints at the enormous cascade effects that would have played out with the extinction of North America's Pleistocene megafauna.[4] Lose the diversifying role of the largest land mammals, replace them with fenced-in cattle grazing the same pastures year-round, and the mosaic of grass, scrub, and woodland becomes invaded by mesquite and white thorn thickets. No wonder the Malpai ranchers fought the U.S. Bureau of Land Management's Smokey the Bear fire-suppression policy: fire, used judiciously, mimics the ecological role of elephants in opening dense thickets and restoring biodiversity.

Lacking the large herbivores of eleven thousand years ago, studies of the cascade effects of keystone species in the Americas have focused on top carnivores. Aldo Leopold came to realize that his wolf-eradication campaign to increase deer numbers for hunters on the Kaibab Plateau on the North Rim of the Grand Canyon led to the destruction of vegetation and, in due course, a dire collapse in deer populations. Though the severity of the crash has since been contested, the restoration of the wolf in North America shows the profound ecological repercussions of losing keystone species.

The eruption of elk and bison in Yellowstone following the nationwide wolf-eradication programs of the 1930s suppressed willow and aspen woodlands, leading to controversial bison- and elk-culling operations to reduce the browsing pressure and restore habitat diversity.[5] A few radical ecologists argued that bringing back the wolf would reestablish the top-down role of predators in controlling herbivores and stimulate vegetation recovery. They would be proved right.

Short of biosynthesizing large American Pleistocene carnivores like the cave lion, saber-toothed tiger, dire wolf, hyena, and cheetah, could smaller living carnivores exert a top-down control of ecosystems, as Robert Paine's keystone hypothesis envisaged?[6] The answers came from the research of John Terborgh and colleagues showing trophic cascade effects on biodiversity with jaguar removal in Central America and increased bird diversity with the removal of coyotes in the sagebrush canyons of San Diego.[7]

The knock-on effects of one species on others underscore three significant forces creating and maintaining biodiversity: the need for large space for megaherbivores to sustain viable populations; the importance of interactions among herbivores, carnivores, vegetation, carrion, decomposers, and other guilds making up an ecological community; and the importance of mobility in enabling species to evade intense competition, predation, disease, climate change, and other perturbations.

How can eastern Africa, Asia, and Latin America conserve large free-ranging wildlife herds and biodiversity in the face of hunting, land privatization, fencing, habitat fragmentation, land degradation, and the extermination of keystone species? How can the ecological minuet and interplay of species be sustained?

The larger the scale of land conservation, the greater the scope for species survival and a self-regulating ecosystem. There is a hitch in scaling up from park to ecosystem and ecosystem to landscape in settled lands, though. The larger the scale, the greater the need to patch together fragmented parcels of land cutting across a welter of jurisdictions and babble of interests.

In Amboseli I ran up against the politics of scale in the early 1970s in expanding the conservation goals from park to ecosystem. Persuading the Maasai to accept wildlife on their land in return for tourism revenues was hard enough, given their deep suspicions of government. But at least the community was cohesive and the leaders few; reaching a consensus was eased by the uniformity of lifestyle and culture among the Maasai. Thirty years on, the problems called for a new brand of conservation. Farmers, ranchers, traders, entrepreneurs, administrators, and young Maasai schooled in a different world than their parents made consensus elusive. The challenge of reaching agreements among the hodgepodge of researchers, conservationists, tour operators, community groups, and government agencies had become as tricky as the United Nations striving for a solution to the conflict in war-torn Central African Republic.

Rural communities across Africa have taken up the cry of democracy since the fall of the Berlin Wall in 1989 and are exerting newfound rights to their traditional lands and use of wildlife. To me, this is progress after decades of repressive governments thwarting the very freedoms Africans gained at independence and conservationists having more say over wildlife than resident communities.

In 2004, frustrated by the push and pull among conservationists, researchers, tour operators, and the Kenya Wildlife Service, I urged seventy community leaders, funded by the African Conservation Centre and African Wildlife Foundation, to draw up a conservation plan for the Amboseli ecosystem. Over the next three years the variegated group worked though reams of information and held dozens of meetings to prepare the Amboseli Ecosystem Management Plan. Co-signed by the Amboseli-Tsavo Group Ranch Conservation Association and Kenya Wildlife Service, the plan set up the Amboseli Ecosystem Trust, an oversight body made up of a group of ranchers, government officials, conservationists, researchers, and tour operators, each assigned specific roles. In 2015 the plan was legally registered and funded by a consortium of organizations.

I use Amboseli to clarify the evolving goals of conservation in response to local and global changes. Even before the Amboseli ecosystem plan was complete, a resurgent Amboseli elephant population spread into the Rift Valley and into Tanzania. The African Conservation Centre, in collaboration with the South Rift Association of Land Owners it launched in 2004, responded by setting up a network of community scouts between Amboseli and Maasai Mara to ensure safe passage for elephants into their former range, abandoned to poachers three decades earlier. The renewed elephant migrations eased the elephant pressure on Amboseli, linked the fragmented park herds into a large, viable population, and added a missing tourist attraction to lodges set up by the Shompole and Olkiramatian community conservancies in the South Rift.

In 2012 a far larger consortium of communities, conservation organizations, researchers, and government agencies formed the Borderland Conservation Initiative (BCI), aiming to win space for elephants and lions across the Kenya-Tanzania boundary. The landscape, ten times the size of Yellowstone and ranging in altitude from the Rift Valley floor to the roof of Kilimanjaro, is large and varied enough to buffer one of the richest assemblages of vertebrates on Earth against climate change. The test for BCI is to prevent land parceling from disrupting ecosystems, blocking wildlife migrations, and triggering a wave of extinctions.

The conservation obstacles are far greater where land is subdivided, habitats are fragmented, and wildlife is confined to isolated pockets. The 19,500-square-kilometer Kruger National Park in South Africa was ringed with fences until recently. In a familiar story, the compressed elephant herds destroyed woodlands and reduced biodiversity. Scientists and managers tried for decades to stabilize the ecosystem by culling thousands of elephants, carnivores, and herbivores, building artificial water holes, and lighting regular fires to control vegetation. After seventy years of losing ground, Kruger acknowledged the futility of managing the park for stability. In 1998 the South African parks authority switched policy to allow natural processes to keep the ecological minuet in play with minimum intervention.[8] Recognizing the ecological straightjacket of the insular park, the Kruger park authorities removed fences to give elephants freedom of movement across the border into Mozambique and onto adjoining ranches engaged in wildlife enterprises.

Patching together land parcels large enough to conserve South Africa's big mammals and biodiversity is a formidable task, but at least Kruger has the original assemblage of species to work with. The task is far harder where the landscape has been transformed by humans and the extinction of keystone herbivores and carnivores has left a skeleton crew of species. How can any semblance of a coevolved, self-regulating ecosystem be restored when the main players are lost? It can't, of course. The best we can do is to restock wildlife refuges with the remnant crew, win back as much land as possible, and mimic the ecological processes and role of keystone-shaping natural ecosystems. In the prairie grasslands big herds of cattle on the move can serve the same role as the missing bison, for example, and in Malpai Borderlands fire is a surrogate for elephants in reducing bush invasion.

Scientific knowledge and land management skills are vital to the restoration of large ecologically functioning landscapes. So too is overcoming the attitude that the only good wolf is a dead wolf. Conservation groups in America, drawing on newfound ecological knowledge and modern sensibilities, are hoping to restock and rewild the Interior West on a grand scale. Rewilding comes in two versions.

The first was put forward by the paleontologist Paul Martin, author of the blitzkrieg theory of the North American Pleistocene extinctions.[9] He proposed importing African elephants, rhinos, lions, and cheetahs as proxies for the megafauna exterminated in North America eleven thousand years earlier. Martin's Pleistocene rewilding ignores the changes in North American ecosystems since the loss of the Pleistocene megafauna. Plants that coevolved with the North American megamammals over millions of years have likely lost many of their defenses since the Pleistocene extinctions. The new mix of plants, riverine woodlands, and wetlands would be a salad bar for the African elephant and flattened faster than the Amboseli woodlands. The African herbivores and carnivores would, in biological parlance, be invasive species no different than prickly pear cactus in Africa, and far more disruptive. Besides, why should the natural legacy that Africa has preserved be used to restore a lost American age in an alien setting?

The second rewilding vision is eminently feasible. The first steps of implementing it have to some degree erased the stain of early twentieth-century government biologists who maintained that "large predatory mammals, destructive of livestock and game, no longer have a place in our advancing civilization."[10] After protracted studies, public outreach, and lengthy negotiations between government agencies, scientists, conservationists, and ranchers, the gray wolf was finally reintroduced to Yellowstone in 1995. The recovery has transformed the Yellowstone landscape and captured the public imagination. Wolf packs have reinstalled the fear of predators in elk and bison herds and created a cascade of ecological changes that restored woody vegetation and landscape diversity.[11]

Biologists have set their sights on a far grander program for wildlife and wildlands in North America, the Yellowstone-to-Yukon (Y2Y) initiative. The Y2Y program brings together a coalition of researchers, planners, landowners, conservationists, and government agencies to connect the 3,000-kilometer stretch of mountains, plains, and valleys spanning the U.S.-Canadian border. The new vision, like the Borderland Conservation Initiative across the Kenya-Tanzania border and the Kruger initiative across the South Africa–Mozambique border, patches together a network of players, interests, and jurisdictions to scale up conservation from parks to ecosystems, landscapes, regions, and continents.

Such large landscape initiatives aim to secure the space and mobility needed to conserve a multitude of species over a wide geographic range. The difference between conserving species out of fancy or all species everywhere out of empathy and rational enlightenment is a measure of how far our knowledge, values, and sentiments have matured in the last half century. Collectively, these new sensibilities alert us to the finiteness and fragility of our planet and the need to clean our dirty nest. Quite how our understanding and sentiments about nature will change in the future is as hard to imagine as it was for Charles Lyell in the nineteenth century to conceive that we might soon be driven by new sensibilities shucking the extermination of other species.

The best hope for biodiversity lies in keeping all conservation options open as both an ecological possibility and a necessity, as Leopold proposed in his famous dictum: "To keep every cog and wheel is the first precaution of intelligent tinkering."[12] Biodiversity in common parlance is taken to mean the sum of all species. As the term implies, *biodiversity* is shorthand for "biological diversity" and a good deal easier on the tongue and ear. For biologists, biodiversity refers not only to the diversity of genes and wealth of species, but also to the role they play in the assemblage of plant and animal communities and the emergent properties and functions of ecosystems. Biodiversity is nature's orchestra, made up of an ensemble of instrumentalists creating a rich symphony of sounds.

The options for conserving biodiversity range from alleviating stress to restoring species and habitats. The conservation tools for intelligent tinkering include rescuing species through captive propagation, restoring them to the wild, and reconstituting ecological processes. Successful examples include the reintroduction of the Arabian white oryx, the genetic and behavioral management leading to the restoration of the golden lion tamarin to the Brazilian Atlantic Forest, reconstituting wetlands in the San Elijo Lagoon in San Diego, and Shirley translocating the Pumphouse baboon troop, which was losing out to immigrant farmers, to a new location.

The oldest tools in the conservation arsenal are hunting regulations and wildlife reserves. These trusty standbys have stemmed the tide of extinctions and saved vast tracts of natural lands. How much biodiversity survives, and how intensively we must manage and restore species assemblages, comes down to our skills at husbanding nature. Keeping every cog and wheel is not enough, unless we wish the wilds to become a theme park. We must also win back the space and restore the natural processes ecosystems need to function, as Kruger National Park is attempting to do.

We are far from understanding the intricacies of ecosystems, much less the biosphere. Scientists are using supercomputers and complex models to track nature's myriad pathways and model the intricate workings of ecosystems. With human activity changing the once-languid pace of evolution and climatic flux from millennia to years, with computer simulations becoming too complex to grasp yet reliant on simplifying assumptions too crude to predict the twists and turns of natural systems, we often turn to metaphors and analogues for our understanding of nature. Our ancestors invoked Zeus and Poseidon, gods who commanded the sky and seas. We still use Mother Nature and her latest incarnation, Gaia, to explain the seeming self-correcting conditions that keep our world in balance suitable for the diversity of life.[13] The problem with the Gaia metaphor, like all metaphors, is that it imbues nature with the intentionality, moods, anger, cruelty, compassion, and wisdom of our own nature.

Then again, using the past as benchmark for conservation today has merit, but only as a reference point, not a prescription. I've worked for decades with paleontologist Kay Behrensmeyer of the Smithsonian, verifying how bones can help us reconstruct Pleistocene fossil assemblages predating our human imprint.[14] Using modern ecosystems as analogues, I try constructing in my mind an age before the emergence of modern humans half a million years ago, a time when giant elephants, hippos, buffaloes, pigs, and baboons roamed Amboseli. I fantasize about restoring the Pleistocene megafauna with Jurassic Park–like genetic wizardry, only to realize there is no longer any place for them. The Human Age we have created is a shadow of the Pleistocene Age of Mammals, which rivaled in its mammalian richness the Cretaceous zenith of the dinosaurs.

The past as a template for the conservation of parks, ecosystems, and biomes is a halcyon trap. Far too much has changed, and space is far too crimped to restore the aboriginal composition of plants and animals, or to manage communities to function as they once did.[15] National parks, our most benevolent gift to nature, have been transformed by climate change over the past century and will change far more in the century ahead. In "How the Parks of Tomorrow Will Be Different" Michelle Nijhuis notes that of the 150 glaciers in Glacier National Park in 1850, only 25 remain today and that intensifying droughts will reduce Joshua Tree National Park's iconic trees to a few remnant pockets by 2100.[16]

The flaw in equating conservation with biodiversity is that biodiversity is too intangible to have the public appeal of the Grand Canyon or Namibia's Skeleton Coast. Biodiversity lacks the charismatic appeal of whales, pandas, bears, wolves, and elephants. The thousands of tiny invisible invertebrates in the Amazon forest don't tug on the heartstrings as do emperor penguins in the Antarctic battling driving blizzards to raise their chicks, or wave after wave of wildebeest churning across the Mara River through rafts of lunging crocs. We are moved more by the sounds, sights, beauty, and drama in nature, by things that stir our emotions, not by abstract numbers of anonymous species.

Biodiversity is a handy catchall term for capturing the wealth of species, some useful, some threatening, others endearing, and many dangerous and loathsome. Mother Nature is unlike our real mothers who nurse and nurture us. For most of human history other species have stood in the way of our survival, growth, spread, and co-option of nature to our own ends. Contrary to James Lovelock's mothering, nurturing nature, Peter Ward, like many Earth scientists, holds to the Medea hypothesis, which sees Darwinian competition causing extinctions down the ages as one species creates conditions inimical to others—and sometimes to its own survival.[17] Biologist Dan Botkin refers to the dynamic tension between competition and cooperation in nature as discordant harmonies.[18]

People the world over still kill and evict troublesome species to make way for cattle, sheep, corn, and rice. Farmers and herders in Africa drive off elephants and lions threatening their families and livelihoods. Americans enjoy TV shows of Africa's wild animals and spare a few dollars to conserve them, only to call on pest controllers to exterminate cockroaches, rats, and termites in their own home and gophers in their backyards. Americans spent an estimated $8 billion eradicating home pests in 2015 and another $10 billion on farm pests and biological extermination, amounting to a hidden war on nature, to say nothing of brush control.[19] In *Nature Wars*, Jim Sterba describes American backyards turning into battlegrounds with the comeback of the deer, alligator, coyote, bear, raccoon, Canada goose, and other species.[20] The same sliding scale of likes and dislikes applies to all people, regardless of wealth, location, or livelihood. We favor beautiful, social, awesome, safe, and empathetic species and shun the lethal, ugly, and alien.

Local nature wars aside, how are we doing in conserving the 10 million or so species worldwide, excluding bacteria and viruses? Agenda 21, the action plan to implement the 1992 Earth Summit Declaration on Environment and Development; the Convention on Biological Diversity, adopted the same year; and the Kyoto Protocol to combat climate change, adopted in 1997, are political milestones on the road to conserving Earth's biological diversity. I consider it an extraordinary feat that the biologists' definition of biodiversity was adopted at the Earth Summit in 1992: "Biodiversity is the sum of all animal and plant species and ecological and physical processes they depend upon and create, as well as the diversity and variability of genes, species, habitats and ecosystems and the ecological complexes of which they are a part."

Billions of dollars have been spent conserving biodiversity in the poorer nations through the Biodiversity Convention's Global Environment initiative. Kenya's Environmental Management and Coordination Act of 1999 is one of dozens of legislative measures taken by developing countries to adopt the principles of the Convention on Biological Diversity. In 1997, as director of the Kenya Wildlife Service, I expanded Kenya's wildlife policies to cover biodiversity countrywide. Convening dozens of local and government agencies, private and communal landowners, and conservation organizations is no walk in the park. We began by mapping a minimum viable conservation area to secure the protected areas and biodiversity, regardless of landholdings. Next, we urged landowners to establish associations large enough to cover the key wildlife migrations and biodiversity. In the final step, the European Union agreed to support a conservation and tourism trust fund to encourage landowner associations and their partners to set up conservancies, biodiversity enterprises, and ecotourism facilities. In 2010 Kenya's new constitution recognized the importance of conserving biodiversity and declared a clean and healthy environment to be the right of every citizen.

For all the political backing, funding, and government rhetoric, the report card on biodiversity is dismal. In 2002 world leaders set out a series of specific goals and targets for conserving biodiversity under the aegis of the Convention on Biological Diversity. A 2010 review of the targets concluded that none had been met, despite a rise in conservation awareness and national commitments.[21] The World Wide Fund for Nature's "Living Planet Report" of 2016, published in collaboration with the Zoological Society of London, concluded on the basis of fourteen thousand well-documented populations that over thirty-seven hundred species of mammals, birds, fish, amphibians, and reptiles had fallen in number by 58 percent between 1970 and 2012, with no sign of a slowdown.[22]

Parks are not insulated from loss in biodiversity. A survey I published with colleagues in 2009 showed wildlife numbers in Kenya's parks falling by a half between 1979 and 2000, the same ratio of losses found outside parks.[23]

Why the continued decline in biodiversity, despite the rising public awareness, increase in protected areas, strong political backing, and financial support?

The 2010 review of the Convention on Biological Diversity targets set in 2002 concluded that the targets were misdirected, aimed at solving the proximal threats such as habitat loss and quick fixes such as protected areas rather than the underlying causes.[24] There needs to be a clear link between biodiversity and ecosystem services, the report concluded. Ecosystem services provide us with food, water, fabrics, shelter, wood products, and medicines, as well as buffer us from environmental hazards; form soils; drive nutrient, geochemical, hydrological, and atmospheric cycles; and feature in our cultural values, including recreation, aesthetics, education, science, and our sense of place and well-being.[25]

Natural capital accounting, the economic valuation of nature's services, such as cleaning up our waste products, recycling nutrients, water catchment, and flood regulation, is still at a rudimentary stage of itemizing, valuing, and quantifying the benefits of biodiversity. University of Cambridge professor Partha Dasgupta, who pioneered the economics of sustainable development, argues that none of the standard measures of the economy are sufficient to account for the wealth of a nation: "Poverty will only be made history when nature enters economic calculations in the same way that buildings, machines and roads do."[26] We should, in other words, include all assets—commodities, infrastructure, investments, institutions, liberties, and social and natural capital—to give a full measure of human well-being—taking into account intergenerational equity and sustainability.

The case for natural capital accounting is growing in proportion to the cost of environmental degradation to human lives and the economy caused by poor resource management. Robert Costanza and colleagues, who have developed full accounting methods, calculate that ecosystem services underwriting our economic progress and well-being amount to $145 trillion a year, twice the global GDP.[27] New accounting methods such as the Economics of Ecosystem Services and Biodiversity show that sound policies, public engagement, and a full valuation of ecosystem services, backed by market incentives, can reduce biodiversity loss and improve human health and well-being.[28] Forests account for a quarter of the carbon sink and offset much of our greenhouse gas emissions. The value of current losses and degradation of forests runs at between $2 trillion and $4.5 trillion a year. Lose the forests and it would take trillions of dollars to restore them and mop up enough carbon to keep global warming within tolerable limits. Preventing the loss of forests can be achieved for an estimated $45 billion, a 100:1 return on expenditures.

Few countries have mapped their biodiversity, let alone valued its wealth of services. Costa Rica and Mexico are exceptions. Both have published stunning illustrated atlases highlighting the country's natural pageantry and its value. In Kenya I led a team of thirty-five government and private-sector contributors in compiling *Kenya's Natural Capital: A Biodiversity Atlas*.[29] Aimed at schoolkids, the business community, and government decision makers, the atlas is a compendium of Kenya's biological wealth and ecological services.

There is, however, a hitch in using natural capital and ecological services to justify conservation of biodiversity. A well-managed timber forest can produce a higher return from a few plantation species and provide many of the ecological functions of a biodiversity-rich natural forest, including water catchment, flood control, carbon sinks, and insulation from solar radiation. The heavily terraced lands of Konso hill farmers in Ethiopia capture water and arrest erosion better than the sparse natural vegetation, yet every species on these lands is domesticated and husbanded.

However large an area we set aside for biodiversity, it will never conserve the natural vegetation that existed at the start of the Neolithic eleven thousand years ago, much less the fragmented and highly modified habitats at the beginning of the twenty-first century. Even if we could, insulating natural assemblages of animals and plants from the planetary-scale changes we are causing is no longer feasible.

The most realistic option for conserving biodiversity on a large scale lies in speeding up the economic and demographic transition: curbing population growth, material consumption, and waste disposal. We also must intensify food and resource production to be as efficient and sustainable as possible and diversify the values we hold for nature to include human well-being. These measures—coupled with the new values, sensibilities, and possibilities ecological emancipation has created for us—offer the green shoots of hope. Meanwhile, we have a far larger and more immediate challenge to confront— how to sustain planetary health within safe boundary limits.[30]

21

The City and the Planet

We are gathered around the campfire on a hunting safari in 1959. High overhead a small dot arcs slowly through the darkening sky. Puzzled by the shooting star that doesn't burn up or fall to Earth, Mel Thomson, the father of my school friends, suggests a Russian satellite. I stare at the receding blip, astonished by the awesome feat of human ingenuity passing over the Itigi Thicket, an impregnable tangle of thorn bush stretching across western Tanganyika.

The human star is the final step of our epic journey of breaking one nature boundary after another. A far longer voyage into space would soon give us the first glimpse of our own small satellite cast against the inky blackness of space. The view of the blue planet fleeced with white clouds taken by the *Apollo* mission in 1968 highlighted the fragility of our Earth far more than thousands of environmental protest marchers, *Silent Spring*, and *The Population Bomb*. Seen by billions around the world, the *Apollo* image took us out of our earthly cocoon and showed us how utterly unique and tenuous is the tiny sphere supporting life in our solar system. The ten-kilometer-deep meniscus of soil and water, sandwiched between the boiling magma beneath and frigid space above, is the fragile domain of all life on Earth. It remains a wonder how a tiny spark of organic life surfaced from the just-right conditions in our otherwise hostile solar system to transform our planet into a world befitting us.

We understood the basic workings of the universe long before we did the origins of life or the functioning of nature. Newton still believed in the biblical creation story when he defined the laws of motion in his *Mathematical Principles* in 1687. Only in the last century and a half have we begun deciphering the great swings and convulsions life has survived since its birth some 4 billion years ago. The miracle of life is that it did survive, adapt, and evolve in the face of boiling heat, a snowball Earth, pyroclastic eruptions, asteroids, and glacial cycles.

In the process of its evolution from the first primordial creatures, life created an insulating atmosphere ever more conducive to further life-forms. Cosmologist Carl Sagan noted that extinction is the rule, survival the exception.[1] He was right in that 99 percent of all species that ever lived are extinct, but he omitted to say that more species are alive today than in any previous geological era. Species have replaced species in a steady clock-like tick of mutations. The fitter and luckier species survived five mass extinctions and rose to a peak of diversity numbering some 10 million species, excluding bacteria and viruses. New species have expanded into vacant niches, and the very diversity of species has created a yet richer community of plants, animals, and microorganisms.

Until we came along. In just eleven thousand years we have captured an outsized share of the sun's energy flowing through the food chain and risen to supreme ecological dominance. By domesticating plants and animals, selective breeding, transforming the landscape, manufacturing nutrients, and exhuming fossil energy, we have boosted production and created a world more conducive to our safety, comfort, convenience, and enjoyment.

The Russian satellite over the Itigi Thicket signified our ambition to conquer space even as the majority in Africa lingered in subsistence economies and wildlands enveloped the continent. The contrast between the two worlds faded when I left Africa for Britain and spent six years in the heartland of the industrial revolution, then lit up again when I settled into Amboseli in 1967 and watched satellites pass over my tent to the grunt of lions and bray of zebras off in the dark. Daunted by the immense journey we had taken from our savanna birthplace to outer space, I had arrived in Amboseli at a pivotal juncture between our ancient and modern worlds.

The swamps fed by aquifers from Kilimanjaro gave Amboseli a feeling of permanence, a place where our ancestors coevolved with savanna wildlife over millions of years. I had no illusion that a national park alone would protect the wide-ranging wildlife herds, but I felt sure at the time that winning Maasai support and securing the seasonal migrations would insulate Amboseli from global changes. I was mistaken. Even as I settled into Amboseli, the glaciers on Kilimanjaro were shrinking, warmed by the effluent of a distant industrial world.

Fifty years on, my hunch in using Amboseli as a microcosm of our African birthplace being drawn into the global age has paid off. Within a few years of starting my monthly monitoring of the ecosystem, the seasonal regularities surfaced from the mass of data and noise of natural oscillations. The clearest signature showed up in the migration of wildebeest, zebra, elephant, and Maasai herds, triggered by the short and long rains, driven in turn by the yet larger migration of the sun passing over the equator twice a year and entraining the storm clouds.

Two decades later the outliers, the extreme wet and dry years, stand out from the regularity of seasons. The erratic years are hitched to the El Niño Southern Oscillation in the eastern equatorial Pacific. Here, every few years, the sea surface temperatures fluctuate from warm in the El Niño phase to cool in La Niña alternation, sending ripples through ocean currents and atmosphere. The fluctuations link remote and seemingly unrelated oceanic and weather patterns around the world. The La Niña cool phase carries a bounty of phytoplankton on the upwelling of the Humboldt Current along the Peruvian coast, giving fishermen a bumper harvest of anchovies. In East Africa La Niña spells poor rains, a dearth of grass, and hard times for herders. These interlinked natural climatic and ecological oscillations can be traced back thousands of years in the growth rings in trees and corals.

The longer I monitor the ecology in Amboseli, the clearer the natural linkages become. The rains set in motion the ecological minuet of wildlife and livestock migrants whirling across the wet-season grazing ranges in pursuit of the erratic storms and greening pastures. The minuet slows and transforms into a grazing procession of animals filing into the swamps in the dry season. At the leading edge of the procession elephants trample down the tall fringing shrubs and swamp sedges, followed by buffalo taking advantage of the clearings. Zebra and wildebeest follow, grazing on the open lawns and shorter regenerating sedges. Thomson's gazelles trail the tail end of the procession to feed on the fresh sprigs of grass far too short to sustain the larger grazers.

Over decades a slower ecological minuet plays out in the gyrating ensemble of animals trekking through the changing kaleidoscope of habitats. Woodlands felled by elephants become grasslands, inviting in zebra, wildebeest, and gazelle. Over centuries a yet more languid animal and plant minuet plays out, triggered by fertile hotspots of abandoned Maasai settlements in which pioneer grasses attracting zebra and wildebeest are replaced by herbs and shrubs attracting Grant's gazelle and impala until, finally, the trees beckon elephants once more.

Over the last fifty years, ecological productivity has become less constrained by the natural forces defining the seasons and carrying capacity, and increasingly driven by the humanizing forces of globalization. After the 1970s drought weakened the erematere link between herders, livestock, and land, pastoral families joined immigrant farmers in slashing, burning, and clearing the fertile wooded slopes of Kilimanjaro for farms, darkening the sky with smoke and cutting off giraffe, eland, and impala from the lushest woodlands. A decade later farmers hacked down the fever trees in Kimana Swamp east of Amboseli to grow irrigated crops, drying up the Lolterish River that feeds the string of wetlands along the base of the Chyulu Hills in a narrow ribbon threading its way into Tsavo West National Park fifty kilometers away.

Had Amboseli not been protected as a park bolstered by community conservation efforts, its swamps too would have been quartered into farms and the springs drained for irrigation. The migration routes would have been cut off and the great herds of wildlife would have disappeared, as they have in most places on Earth with our rise to global dominance.

The African savannas, like the American prairies, have shrunk with the spread of farms, roads, permanent homes, villages, and towns. Grazing pressure due to migratory livestock and wildlife herds squeezed for space has increased sharply across the rangelands of Eastern Africa, shrinking ground cover, increasing rainfall runoff, and reducing pasture production. The worst-hit area, the Horn of Africa, mirrors the Dust Bowl story of Oklahoma and the Texas Panhandle in the 1930s. Too much grass has been stripped from the land as populations grow and settle and trees are hacked down for charcoal, depleting the insulating cover of vegetation. Soil temperatures have risen with the spread of bare ground, and the ability of plants and soils to absorb CO_2 has fallen. Overall, the degradation across the Horn of Africa has further reduced the Earth's capacity to absorb our greenhouse gas emissions.

The scale of human activity has altered every habitat and planetary process.[2] The glaciers of high altitudes and latitudes are the most sensitive and visible markers of global warming. On the Tibetan plateau in 2012 I photographed the glaciers of the Kunlun Mountains, which George Schaller had watched recede hundreds of meters over the past three decades. The polar ice caps have also contracted at an alarming rate. The polar sea ice in the Arctic has shrunk to 2.1 million square kilometers, half its size in the 1980s. Satellite images in 2017 revealed the Western Antarctic Ice Sheet melting and slipping seaward. Computer models based on the new findings project global sea levels will rise a meter or more by 2100 if greenhouse gas emissions continue at the present rate, double the previous estimates and enough to cause severe flooding in most coastal cities. By 2500 sea levels could rise 70 meters if all the ice caps melted. With 2014 the warmest year on record, 2015 warmer again, and July 2019 the warmest month ever recorded, the likelihood of polar ice melting far faster than predicted by the IPCC is in the cards.

As the oceans rise with heat expansion and glacial melt and acidify due to chemical reactions caused by extra CO_2 absorption, coral reefs are bleaching around the world. In 2016 22 percent of the Great Barrier Reef corals, blanched by record sea temperatures, looked like a white stone desert. Like the pasture losses due to heavy grazing in Amboseli, coral bleaching is recurring more frequently and persistently, shortening the window of recovery.

Hundreds of specialists working with the Intergovernmental Panel on Climate Change have tracked atmospheric and oceanic monitoring stations around the world and modeled the impact of greenhouse gas emissions with increasing resolution. At first, the complexity of the interlinked atmospheric, oceanic, and terrestrial systems was too poorly understood and the volumes, sources, and sinks of greenhouse gases too tentatively known for the first-generation computer models to distinguish the human footprint from natural fluxes with any certainty. By 2016 the technical advances on all fronts, along with refined computer models, left little doubt that human activity is warming the planet far faster than natural causes any time in the past few million years.

Just as it has taken thousands of climatologists and Earth scientists to detect and model climate change, an industry of biologists has been measuring, monitoring, experimentally testing, and modeling the responses of thousands of species to climate change. Animals and plants have adapted to climatic oscillations in the past, of course, including cold glacial and warm interglacial periods. So too have we humans over the last half million years since the emergence of our modern ancestors.

The difference in human-induced from natural climatic oscillations lies in the speed of change, and in the human obstacles to plant and animal compensatory movements. The ability of a species to respond fast enough to global warming—to remain within its environmental envelope as its range shifts in altitude and latitude—is hard enough. Add a landscape broken up by farms, roads, and cities, and geographical adjustments to climate change become a mission impossible for many species.[3]

The challenges of climate change and human obstacles show up most clearly in ecological and behavioral changes in high-latitude species. Lapland's million-strong reindeer herd has fallen by a third with the stress of trekking to its northward-shifting foraging grounds and having to browse in less productive higher terrain. Polar bears hunting for seals retreating from the mainland with the breakup of the Arctic ice are having to make longer foraging trips. The added commuting time makes it hard for the bears to build up fat reserves for winter hibernation and suckling cubs.

Funds have poured into climate change research for field studies, lab experiments, metadata mining and analysis, and refined computer modeling. The funding has overcome the aversion of biologists to studying human ecology and behavior and highlighted our pervasive role in every ecosystem. Runaway climate change in the future could be disastrous, but over the last century we have had a far greater impact than at any time in the past in converting habitats to human use and weakening the ecosystem processes that buffer climate. Tim Newbold and colleagues estimate that we have already converted 58 percent of the terrestrial realm to human use and pushed two-thirds of the world's major vegetation zones below recovery levels.[4] Unless we re-create the landscape linkages, many species will go extinct and others will face range constriction, habitat fragmentation, and ecosystem degradation.

The synergies of land use and climate show up in the joint U.S.-Canadian Arctic Program for Regional and International Shorebird Monitoring. The program, which began in 2002, sent dozens of researchers to study the breeding success of twenty-six species of shorebirds at two thousand summer nesting sites stretching from Alaska to eastern Canada.[5] The study found most shorebird populations in severe decline due to a combination of earlier spring warming in the Arctic and land-use changes along their flyways as far south as the southern tip of Argentina. The red knot has suffered a 75 percent decline since 1980 and is now listed as an endangered species. The knot's main refueling depot during its seventy-seven-hundred-kilometer migration north is the energy-rich spew of eggs from horseshoe crabs spawning in Delaware Bay. Fishermen overharvesting the crab have severely reduced the knot's fuel supply. The earlier spring emergence of insects in the warming Arctic also means the knot is missing the peak eruption of its food supply and suffering a lower breeding success and a slower growth rate in chicks.

The threats to the global commons are so diffuse, complex, and slow acting, and have such deep cascading and long-lasting effects, that it is hard to grasp and communicate them in a graphic way. Our global impact is beyond our sensory reach, emotional ambit, and national jurisdictions. We can't rely simply on our conventional evolutionary tool kit or our current knowledge, institutions, and conservation skills to solve our global Dirty Nest Syndrome. The tragedy unfolding in the global commons lacks the proximity effects that triggered action within communities and our national political institutions. The global and generational scales of the twenty-first-century commons lie beyond free-market, command-and-control, and voluntary conservation solutions. The tyranny of two hundred nations, millions of corporations, and nearly 8 billon individual self-interests vying in the global commons would have been Garrett Hardin's worst tragedy of the commons nightmare.

In *Governing the Commons*, Elinor Ostrom shows that a many-tiered approach is needed to solve the planetary tragedy of the commons.[6] It will take the local efforts of Maasai herders and Malpai ranchers, the initiatives of cities, private and public institutions, small businesses, and big corporations as well as a raft of new incentives, policies, regulations, oversight, and enforcement of nation-states, regional bodies, and international agencies to solve the tragedies of the global commons.

The Maasai and elephants of Amboseli are as subject to global forces as they are to local ones in today's hyper-populated world of intersecting markets and social networks. The Chinese ivory market is as likely to decide the future of the elephant as conflict with local farmers. Google, Microsoft, and Apple, more than elders and clans, determine the social networks of young Maasai today.

Fixing the tragedy of the global commons comes down to the same solutions as fixing the tragedy of the local commons. We share the same human nature as our Late Pleistocene forebears and all peoples and cultures around the world. Meeting the global challenge calls for the same collaborative skills that propelled us to superdominance: the feedbacks between action and consequences at every level from household to global institutions. It also calls for raising the productivity and efficiency of our food production, manufacture, and services, uncoupling the links between economic growth and environmental impact, and closing the gap between the developed and developing nations.

The steps I've outlined in the preceding chapters show how we can move up the ladder of concern from our own survival and self-interest to include other people, other species, the environment, and the future of our planet. These steps offer the prospects—not a guarantee—of avoiding a new Malthusian trap, lowering our human footprint, staying within ecological and planetary boundaries, and giving us an opportunity to enjoy nature for the many values it offers us in the global age. More than at any other time in our short and meteoric evolutionary rise, managing our future depends on reconnecting the human and natural world we have spent centuries separating.

Forests have been described as the lungs of the Earth. They absorb about a quarter of the annual greenhouse gas emissions. Rangelands, which cover a third of the Earth's surface, absorb another 10 to 15 percent, mainly through root storage. Hacking down the forests and destroying the rangelands pumps a surge of soot and CO_2 into the atmosphere and lowers our capacity to recycle emissions through natural ecosystem processes. Land clearance and fossil fuel emissions create a self-reinforcing cycle of higher global temperatures, less carbon absorption, and more extreme weather patterns.

Conserving and restoring the world's forests, woodlands, and wetlands to a functioning ecological state is a win-win solution for combating climate change locally, nationally, and globally. Added vegetation cover absorbs greenhouse gases, saves biodiversity, and sustains the ecological services we depend on, including nitrogen, hydrological, and other ecological cycles.

Small steps taken locally can raise public awareness of the part habitat conservation and restoration can play in moderating greenhouse emissions. Wangari Maathai's Green Belt Movement started in Kenya and spread around the world as a voluntary grassroots initiative to replant trees, restore habitats, and rekindle ecological functions and services. The Malpai and Maasai exchanges have triggered a similar movement to restore the health and resilience of the rangelands, sustain cattle cultures, and conserve wildlife.

Safeguarding planetary health is not like managing a well-run family ranch or farm, though. We have no feel for planetary health in the way we do a family dairy herd or vegetable garden. Calculating the human impact on global warming is as complex and arcane as science gets. We can, however, use simplified metrics of carbon emissions to track our ecological footprint. Carbon is the building block of life. We can measure its flow and flux through an ecosystem and the economy. In natural ecosystems and biocultures, energy from sunlight, water catchment, and nutrients determined the productivity of plants and animals. Human population size was limited by the natural productivity of the land. We lived on the annual interest of biomass stored in animals and plants. Our footprint was neutral: the carbon we captured, wild or domestic, built and sustained our bodies and was returned to earth as surely as we died.

Our impact since the industrial revolution and the exhumation of billions of years of carbon stored in fossil fuels has raised the CO_2 level past the four hundred parts per million level and will push up global temperatures two degrees centigrade or more by 2050, unless we take drastic action, and six degrees by the end of the century—far too hot for our present habitation.[7]

We can measure our carbon emissions and track our environmental impact with some precision at an individual, national, and global level and take aversive action in every aspect of human life. The American Institute of Architects, for example, has challenged the architectural community worldwide to become carbon neutral in building designs by 2030. The international Carbon Neutral Cities Alliance aims to show that carbon emissions are a design failure that can be rectified by slashing greenhouse gas emissions 80 percent by 2050.[8]

Time is not on our side when it comes to carbon emissions and global warming. We need decisive action at a global level to prevent temperatures soaring through the two-degree-centigrade threshold Earth scientists consider a break point beyond which runaway warming will take centuries to reverse. The quickest win is to scrap fossil fuel subsidies and invest the savings in clean and renewable energy. Perverse subsidies distort prices, perpetuate flagging fossil fuel markets, and cater to powerful political interests.[9] Tax breaks for fossil fuel companies and oil depletion allowances in the G20 industrial nations cost every citizen $1,000 annually. Add to that cost the expense of health care caused by pollution, and the invisible cost of fossil fuels is the devil in the darkness.

The G20 nations have made a commitment to gradually phase out all fossil fuel subsidies.[10] The IMF estimates that raising the cost of fossil fuels by cutting subsidies would favor the use of renewable energy alternatives, cut CO_2 emissions by 20 percent and, in the process, raise government revenues by $2.9 trillion, or 3.6 percent of GDP worldwide. The savings could be invested in the research and development of clean fuels and pay to protect and restore habitat cover, biodiversity, and the ecological services they supply us.

Fifty years on from Garrett Hardin's dismal rendition of the tragedy of the commons, we know a good deal about the principles of collective action that work locally and nationally in managing public utilities like parking lots, highways, and airwaves. The principles fail to prevent overfishing in the open oceans and greenhouse gas emissions in the atmosphere where governance is lacking. Managing the global commons calls for levels of collaboration, knowledge, and jurisdiction far beyond national boundaries and the political cycle of any nation. Several global agreements reached in the twentieth century have shown success in, for example, closing the ozone hole under the Montreal Protocol and saving crocodiles and spotted cats from the international wildlife trade under the Convention on International Trade in Endangered Species. So why have we failed so dismally to curb greenhouse gas emissions, which have far greater global consequences?

One reason is that we don't feel the consequences of the slow warming of the planet, acidifying oceans, and melting of the glaciers in the way we do the city smog and fouled rivers that directly affect our health and quality of life. The polluted rivers and skies of the United States in the 1960s sparked the environmental movement, street marchers, and public condemnation of the biggest polluters, leading to the establishment of the Environmental Protection Agency in 1970. Similar movements are growing in China, India, and other nations as the polluted skies and rivers make life unbearable for the rich as well as the poor in the industrializing heartlands and congested cities of the emerging economic powers.

We have yet to establish the international trust and governance mechanisms to manage remote global disruptions. The lengthy negotiations following the Kyoto Protocol on reducing greenhouse gases adopted in 1997 illustrate the problems. After a decade of talks, negotiations broke down acrimoniously at the Copenhagen Conference of Parties in 2009. Several factors contributed to the wrangling, among them disagreement over the historical role of the West as the biggest greenhouse gas emitter, its responsibility for compensating the rest of the world now suffering the consequences, the level of reductions, the share expected of each nation, whether the targets should be based on total emissions or emissions per capita, the need for technological transfer, and the like. The sluggish global economy in the wake of the Great Recession of 2008, falling emission levels, and plunging oil prices at the time also played a role.

The contrasting success of the Paris Agreement reached in 2016 was credited to a new IPCC climate change report giving stronger evidence of global warming. The signals were also being felt in daily life in the spate of fierce hurricanes, floods, heat waves, the rapid breaking up of Greenland glaciers, and sea-level rises flooding low-lying Pacific islands. NGO and public pressure groups, Pope Francis's encyclical on environmental stewardship, and a bilateral agreement between presidents Barack Obama of the United States and Xi Jinping of China pledging to reduce the emissions of the two countries—all added to a groundswell for an accord. The combined pressure led to lengthy dialogue and voluntary yet audited measures to cut emissions to keep global temperature rises within two degrees centigrade.

Nobel laureate William Nordhaus of Yale University calls for climate clubs—companies and groups sharing a common commitment and values, creating incentives to voluntarily reduce carbon emissions, and imposing penalties and trade barriers on defiant nations and companies.[11] Nordhaus sees climate clubs as creating a virtuous cycle, drawing in more players who benefit from cleaner air and lower energy consumption brought about by greater efficiency in resource use. The moratorium on whaling issued by the International Whaling Commission in 1946 and the ivory trading ban by CITES in 1989 resulted from just such public lobbying and peer pressure, sparked by the outpouring of sympathy for harpooned whales thrashing around in bloodied seas and elephants with their faces hacked off by ivory poachers.

Greenhouse gas emissions lack the emotional valence of slaughtered whales and elephants and seem far too remote and nebulous to serve as a rallying point for public pressure or climate clubs. And yet a coalition of hundreds of mayors from cities across the continents claiming to represent half the world's population played a pivotal role in steering the Paris Agreement to a final agreement in 2016. Why, of all places, should cities become a rallying point for curbing climate change, and what role can they continue to play?

Cities are the terminus of the global express that started in the African railhead half a million years earlier. Oddly, despite bearing no resemblance to the savannas, the city is our new ecosystem recapitulating the conditions that made us the dominant and always adaptive species we are. Over half the world's population lives in cities, up from 30 percent in 1950. Three-quarters of us will crowd into cities and occupy less than 5 percent of the Earth's surface by 2050. By then cities will be draining the rural populations in the developing world in a reprise of the efflux in Europe and America since the start of the industrial revolution. China, nearing the end of the greatest migration in human history, has grown from 20 to 50 percent urban in thirty-five years. The rural exodus in India, Southeast Asia, and Africa will be well advanced by the turn of the century.

Humans in cities defy the tendency of other densely crowded species to slow growth in response to shrinking food supplies and contagious diseases. Cities are the origins of our civilizations, the centers of power, the birthplace of modern industrial states, and the laboratories of innovation.[12] Despite higher crime rates and stress, cities are magnets for young people like Josh Kirinkol who seek a richer life, and others fleeing poverty and land shortages.

In a surprising twist, cities follow Kleiber's law of metabolic rate in relation to body size noted in chapter 9: for every doubling in size, energy efficiency increases by 15 percent.[13] We have already seen human societies from hunter-gatherers to agrarian societies become more efficient in larger groupings, and cities are even more efficient.[14] Just as the length of arteries, veins, and capillaries supplying energy to animal cells decreases in proportion to body size, so too the length of roads, electricity cables, and water and sewage pipelines decreases per capita with city size.[15] The bigger the city, the cheaper the per capita costs of utility installation and maintenance costs as well as the provision of schools, hospitals, libraries, communications networks, and other public services.[16] Most urbanites use public rather than private transport, and water and trash are recycled more efficiently than in small towns.

The efficiency, amenities, and services of big cities attract and spawn businesses, create jobs that offer higher wages, and provide better health care, denser communications networks, and greater creativity in the economy, arts, sciences, and entertainment. I am no city person but admit to finding the throb and pace of life energizing in New York, Los Angeles, London, Paris, and Beijing whenever I visit. A short subway or Uber ride opens a world within a world—museums, art galleries, theaters, restaurants serving cuisine from every quarter of the world, stores offering a cornucopia of goods and services, and a coterie of cultures drawn to the cosmopolitan atmosphere of city life.

Cities grow on average at a super-linear rate of 1.15, meaning that doubling the size of a city more than doubles the social and economic opportunities.[17] The pace of life is altogether faster and more productive in big cities. People even walk faster than in rural areas. In two fundamental respects, though, the metabolic power law differs from that of organisms. As a species we each burn the equivalent of a ninety-watt light bulb of energy each day, about half as much again as a baboon. Yet the total energy we burn in warming our houses in winter and cooling them in summer, in building our homes, driving our vehicles, growing our crops, feeding our animals, and transporting commodities around the world amounts to eleven thousand watts, the same as an elephant burns each day.[18] Multiply our individual energy output by the total world's population and the sum totals nearly 100 billion elephants' worth. This astonishing figure graphically illustrates our global ecological footprint. If fewer than a thousand elephants crowded into Amboseli can destroy woodland and disrupt biodiversity, imagine the impact of 100 billion elephants.

The analogy between cities and organisms breaks down in the case of self-limiting growth. Unlike organisms, cities are not closed systems limited by energy. Cities thrive on growth and feed on the innovations driving economic productivity and efficiency.[19] The super-linear growth of cities sucks in people and pumps out effluents. The fossil fuels powering cities give the illusion of infinite growth and the delusion of cost-free waste. After all, Los Angeles, New York, and other Western metropolises have become clearer in the last few decades despite continued growth. Global warming brings us back to the reality of a closed planet and the limits to growth. The limits are less about the energy powering growth, given the prospects of renewable solar, wind, and wave sources, than about the ability of our planet to absorb pollutants and function within the narrow ecological envelope in which we thrive, and the diversity of life depends.

Cities are the engines of the twentieth-century economy and the main source of new ideas, new technologies, and global connectivity.[20] Cities speed up the demographic transition and are also uniquely cosmopolitan; as meeting places of peoples from different ancestries, religions, and culture, they create a convergence of interests and aspirations after millennia of divergence.

Cities, one way or another, will define the future more than any other of our creations.[21] They reflect the best of us and the worst of us. They recapitulate our roots in close-knit communities responsive to their social and environmental impact and dirty nesting habits. The environmental movement that began in the polluted cities of the West is playing out again in the suffocating cities of China and India. Is it any surprise, then, that mayors around the world are forming a global coalition to make the city a cleaner, more productive, and more efficient place to live and do business? The fate of our cities will define the fate of our planet.

Conclusion: From Conservation Philosophy to Citizen Action

I've given a broad sketch of the role of conservation in our struggle to survive and outcompete other species, its expansion with the domestication and husbandry of crops, livestock, and land and, ultimately, its culmination in conserving nature, culture, and knowledge in our rise to global domination. Living within the limits of our planet in the Human Age is the ultimate test of whether the conservation ethos is deeply enough ingrained in society for us to create a world we would wish to inherit.

My own life has followed a similar trajectory to the diversification of conservation, from the thrill of hunting wild animals for the pot to the far greater satisfaction of saving species. Growing up among traditional communities, I came to appreciate how my passion for wildlife arose out of my ecological emancipation from the daily hardships my distant ancestors faced in making a living.

Having one foot in Africa and another in California has given me a rear view of our Paleolithic birthplace and a front-row seat in our human-built world. Metaphorically, our global domination reenacts the closing of the seemingly endless American frontier in the nineteenth century, revealing Earth's finiteness and vulnerability to our conquest. Wallace Stegner's reproach that cooperation must replace rugged individualism to save and restore the West applies no less to politicians, corporations, and peoples whose bottomless cookie jar view of Earth must give way to global collaboration if we are to sustain the health of the planet and the richness of life.

Our past is littered with wildlife skeletons, razed forests, eroded soils, and abandoned civilizations. Ancient societies failing to live within local limits could move on and nature could recover. Knocking the biosphere out of kilter is another matter altogether. Primitive life took billions of years to oxygenate the atmosphere to make it suitable for complex plants and animals to evolve, another billion for the diversity of life to reach an azimuth, and a mere third of a million for us to emerge as *Homo sapiens* and colonize the world. In just three centuries more we have risen to a supreme dominance, rupturing cultures and the diversity of life. Destabilizing the planetary homeostasis is akin to knocking our metabolism out of whack with an overdose of drugs.

Decades of philosophical and academic research into how to expand the ethical treatment of one another, the land, other species, and nature have floundered. This is no surprise: we have trouble enough getting along with cultures and tribes different from us—let alone locusts, termites, tigers, and elephants ravaging our life and property.

We've made some progress in developing a body of science for managing natural resources such as forestry, fisheries, and watershed management, and in reducing pollutants directly imperiling and impairing our lives. Where we fail is in putting our knowledge into practice in the global commons because we are not that rational actor of economic theory making the best possible choices. We are fallible, emotional beings with competing interests. We screw up for all sorts of reasons yet do manage now and then to rise above the Machiavellian fray to achieve greater things for the common good, despite our foibles. Here lie lessons for managing the global commons.

I've found it useful to look at the underlying causes of conservation successes and failures to explain why we sometimes manage to conserve things we value yet so often fall short. Conservation is not unique to us. Other species are evolutionarily primed to survive harsh times, from polar bears storing body fat and hibernating to survive the Antarctic winter to desert tortoises storing water in their urinary bladder to evade droughts. Like polar bears, we too store body fat in good times and turn down our metabolic thermostat in hard times to conserve energy. Yet it took something other than an in-built biological survival kit for us to outcompete the large herbivores and fierce carnivores in the savannas.

The hallmark of our success lay in breaking the evolutionary constraints of our mammalian body plan and kinship bonds, taking stock of our environment, anticipating shortfalls, manufacturing tools, and planning and coordinating hunts among large, often unrelated groups. We pressed home our ecological advantage and social skills by domesticating, biologically engineering, and husbanding plants and animals to produce our own food and supplies. Domestication gave agrarian populations an ecological advantage over hunter-gatherers and spread like tumbleweed around the world. The genes, languages, customs, and cosmologies coevolving with our husbandry practices created new cultures able to solve the tragedy of the commons and tyranny of small decisions. Early farmers and herders cooperated to mutual advantage to raise and sustain productivity by mimicking natural energy flows and nutrient cycles. The Konso Highland farmers of Ethiopia, for example, have husbanded their rich farming system for centuries by terracing steep rocky hillsides to slow erosion and harvest water.

Our global superdominance has melded and molded economies, markets, lifestyles, cultures, and governments after millennia of diversification at an ever-quickening pace. In half a generation I watched the Maasai warrior-spokesman Kerenkol Ole Musa transition from subsistence cattle herder to commercial farmer, taxi operator, and lodge owner merging into the new Kenya nation and global economy. In a mere quarter generation more, I watched his son Josh win a degree in Switzerland and take up work in the hospitality industry on the ski slopes of Aspen.

The great convergence and homogenization of the global age has freed most of the world from farming, herding, and producing as many children as possible. We can now enjoy nature for many reasons—recreation, adventure, discovery, spiritual solace, education and, for the likes of me, the joy of watching wildlife and discovering the workings of nature. Billions of people still locked in poverty could enjoy freer, richer, safer lives, if given the opportunity.

Superdominance heralds a new evolutionary age of our own making, the Human Age. We have thrown out the selfish gene and an environmental straitjacket constraining us biologically. We have transformed wildlands to farms, ranches, and human settlement. We have hacked down forests, drained lakes, dammed rivers, and ripped apart the Earth for oil and minerals. Our own nature has changed in our human-built world, from an intimate and lowly member of the food web to a novel species creating our own niche and conserving the very habitats and species we once destroyed. The by-products of our profligacy are sullying our own nest to the point of heating up the planet, polluting rivers, acidifying the oceans, poisoning soils, and diminishing our own lives.

We have become a new sort of species—at once a legatee of our primate ancestry and a turbo-charged novel biological force reshaping the planet and driving evolution. No species can avoid the new selective forces of our superdominance and landscape engineering. Songbirds in New York call louder to attract mates and drive off rivals above the roar of traffic. Lions fall silent to avoid detection when they move into Nairobi suburbs to prey on warthogs grown fat on green garden lawns more enticing than the dry grasses of Nairobi National Park. On our land next to the park a lioness, eyeing our yard as a haven, cached her cubs outside the back door and went off hunting. Shirley and I welcomed the visits until another lioness took our generosity too far, helped herself to one of our dogs, and spent three days devouring it in our yard.

The schism between natural and human-made worlds is narrowing, and perhaps was never more than a temporary phase in our shift from rural to urban living. In an odd twist, we are repeating the nature wars of old. Instead of us invading nature, foxes in Europe, coyotes in suburban America, and macaque and langur monkeys in Delhi inveigle their way into our cities.

We have become omnipresent. Some claim we have become omniscient too, replacing the gods in correcting and redesigning the faulty and inelegant blueprints and body plan of our genes, hormones, anatomy, physiology, and intelligence. In a new evolutionary twist, we are on the threshold of resurrecting species we long ago exterminated and creating entirely new ones to suit our purpose and fancy.

Such hubris needs a reality check. Our knowledge of ecological and planetary science is paltry alongside our grasp of economics, yet if we failed to predict the global financial meltdown of 2008, how can we risk centuries of disruption if our ecological tinkering proves as fallible as the financial quants whose fancy equations ruined the lives of millions? The convulsive disruptions of the COVID-19 pandemic on billions of lives in

2020 proved how ill-prepared we are for global calamities, and how misplaced our hubris.

There is nothing intrinsic about our caring for the environment or other species. We have no gene for saving nature and no inbuilt feedback allowing us to see, feel, and counter the global mess we're causing. We are neither the economist's rational actor nor the biologist's selfish gene and blind watchmaker. Rather, our human nature includes the ancient self-interest of all species, overlaid by our unique capacity to cooperate for mutual long-term gains. A safe transition from ecosystem to planetary limits depends on exorcising our blind genes and exercising our foresight and cooperation on a global scale.

The crucial first step is to reconnect our food supply chain: from farm to plate and from factory to consumer, reusing resources as if we lived in a small, isolated village. We can't of course forge global connections using our natural senses, but we can use extrasomatic proxies like the internet and social media to monitor the distant consequences of our actions and right harmful damage. Once we can see and sense our distant impact, we can take responsibility for our actions, starting with the things we depend on most—food, water, and energy. We can speed up progress by uncoupling the link between economic growth and environmental impact, closing the gaps between poor and rich nations, and accelerating the demographic and economic transitions as outlined in chapter 16.

These steps can help us avoid the Malthusian trap of populations relentlessly growing to the limits of food availability, lower our human footprint, keep us within ecological and planetary boundaries, and allow us to enjoy nature. Freed of the day-to-day labor of producing our own food and supplies, we can expand our circle of concerns to include environmental health, security, and our own well-being. These steps, adopted by nation-states and regional economic blocks such as the EU, NAFTA, AU, and ASEAN, show up in health, social, and environmental standards among trading nations.

Conservation has made progress over the last century in conserving soils, watersheds, and forests; reducing pollution locally; and saving the bison, panda, and whales. The global threats in the twenty-first century are still mounting, though, and are so ubiquitous and deep-seated as to defy the efforts of any one nation. The greatest challenges are global warming and the disruption of planetary cycles, but ultimately all solutions come down to uncoupling the link between GDP and resource consumption.

We have evidence that the link can be broken when resources are scarce. Had the United States continued its 1980s level of per capita water use, consumption would be 40 percent higher today. The same can be said of pollution. If not for the environmental movement, Americans would by now be choking in smog far worse than Beijing's acrid pall.

On the upside, resource efficiency—the amount of resources we use per unit of economic output—has risen by over a third and waste efficiency by nearly 50 percent worldwide in the last decade. On the downside, the gains have been diluted by the absolute growth in production and waste. The newly industrialized and emerging nations will need to leapfrog the smokestack technologies of the West to avoid a ruinous growth first, clean up later path to development.

The Paris Agreement, signed by 195 nations in 2016, is a remarkable testament to the rise of the climate change crisis to the top of the international agenda. The accord scored a victory for global collaboration after a quarter of a century of self-interest, blame, compensation claims, and insistence on fixed targets and timetables scuttling earlier agreements. The change in the political tone stemmed in part from the extensive consultations ahead of the congress and better science informing the world delegations. The most compelling factor, however, was public pressure, driven by the visible signs of warming brought home by sea-level rise drowning Pacific islands, Hurricane Sandy devastating the eastern seaboard of the United States, and glacial melt in the Himalayas causing severe floods in downstream farms and towns. Seeing and feeling the impact of global warming was worth a thousand graphs.

Global cooperation is still threatened by regressive political shortsightedness, populism, and religious and national fundamentalism. The election of Donald Trump as the U.S. president and his "Make America Great Again" call to revive the mythical halcyon age of coal power and heavy industry have put global cooperation to the test. H. R. McMaster, national security advisor to the president, and economic advisor Gary Cohen stated the new U.S. policy in the *Wall Street Journal:* "The world is not a global community but an area where nations, nongovernmental actors, and businesses engage and compete for advantage. Rather than deny this elemental nature of international affairs, we embrace it."[1] In line with the new policy, Trump announced the United States' withdrawal from the Paris agreement.

Two decades ago, the Paris Agreement would have stalled for lack of U.S. leadership, given America's global dominance. In the twenty-first-century networked age of greater awareness of global threats, cooperation no longer relies on the suasion of a single superpower. Michael Bloomberg, former mayor of New York City, responded to the U.S. plans to exit the Paris Agreement by pointing out that the global commitment to reduce carbon emissions will continue with or without America's leadership.[2] Consumer demand for cleaner energy and cheaper alternatives is pushing over half of America's coal companies to close coal mines and invest in renewables. Wind power is displacing coal-fired production as the cost per megawatt has dropped to $20 compared to $30 for coal-fired energy. More than 130 U.S. cities have joined the Global Covenant of Mayors for Climate and Energy in line with the Paris Agreement. Voluntary collaborative action can fill the void left by a U.S. withdrawal and meet the Paris Agreement goal of reducing greenhouse emissions by 26–28 percent of their 2005 levels by 2025. The reductions can be achieved by a combination of market forces, energy efficiency in vehicles and utilities, and the unilateral commitment of cities to reduce emissions, according to a *New York Times* analysis.[3]

The defense of the Paris Agreement shows that we respond best to environmental threats when the consequences of our actions are visible and give us a sense of engagement and responsibility—when we can play our part and take pride in doing the right thing, and when our actions are subject to public scrutiny. In the globally connected age, a change in cultural expectations, norms, and practices could happen faster than at any time in history and become as natural as handwashing.

Sustaining a healthy biosphere also comes down to conserving species and their role in maintaining life processes, including our own. The task, no less daunting than curbing climate change, calls for the same deep understanding of natural and human processes and cooperative solutions, as the following example illustrates.

When my wife Shirley first studied the Pumphouse Gang troop of baboons on Kekopey Ranch in the Rift Valley in 1972, the faulted scarps descending from the Mau Forest to the checkerboard of grasslands, *leleshwa* scrub, and fever tree woodlands on the rift floor looked much as it would have done to our hunter-gatherer forebears in the Pleistocene—apart from cattle scattered among the herds of wildlife. Baboons still foraged across the ranch in an endless daily cycle of feeding, socializing, mating, and sleeping on cliffs at night, posing no threat to the rancher, Arthur Cole. Yet for all the seeming continuity of baboon life, Kekopey epitomized our changing views of human and animal rights in the global age, and the very notion of natural and humanized landscapes.

Arthur Cole's rights to private land were being contested at the time by the Maasai, whom the British colonial government had evicted fifty years earlier to make way for white settlers. The Maasai themselves had defeated other pastoral tribes in the Rift Valley five hundred years beforehand, and yet earlier pastoralists had subordinated indigenous hunter-gatherers a few thousand years earlier. The few remaining hunter-gatherers, the Okiek, were confined to the Mau Forest and threatened with eviction by illegal loggers and powerful land grabbers. Highland farmers from the dominant tribe, the Kikuyu, set up a land-buying consortium, GEMA, which bought out Arthur Cole in 1978, parceled up the land, fenced off individual plots, and converted ranch land to fields of maize.

Zea mays (called corn in American English), native to Central America, is an annual grass transformed into a super crop rich in energy and starch from wild varieties some eight thousand years ago. Introduced into East Africa around the sixteenth century, maize rapidly spread across the continent and today provides 30 percent of the calories consumed in Africa.[4] Once the farmers began planting on Kekopey, the Pumphouse Gang took to raiding the maize cobs (far richer in starches and more digestible than native grasses), blind to property rights and oblivious to the distinction between natural and human-modified foods. The crop raiders, like fast-food junkies, put on weight, matured faster, and produced more offspring than neighboring naturally foraging troops.

Farmers, angered at their crop losses and by baboons raiding their homes, took up arms and killed several baboons. Shirley, like thousands of other researchers around the world, became a conservationist to save her study animals. To begin with, the solution seemed straightforward enough: hire chasers to protect the farmers' crops and spare baboon lives. Meanwhile, Eburru, an adjacent baboon troop wary of farmers, continued feeding on the native grasses, lost land to settlers, and shifted its daily ranging.

Whose rights to the land are paramount at Kekopey—those of the baboons, whose presence preceded humans; the Okiek; the Maasai; Cole; or the farmers with newly issued title deeds? And what is the natural thing for the Pumphouse baboons to do, feed on maize to boost their reproductive success, or continue feeding on native grasses and lose condition and range, as the Eburru troop did?

Unable to keep the Pumphouse Gang out of trouble, Shirley took a drastic step and translocated three troops of baboons north to Chololo on the Laikipia Plateau 150 kilometers to the north. Here she paid the rancher a research fee and hired field assistants from the adjoining community to give them a stake in conserving and studying baboons. She later raised funds to build a boarding school and help the kids of poor pastoralist families stay in class and stand a better chance of finding jobs in Kenya's competitive job market. The baboons quickly adapted to the drier environment and began killing goats. Shirley then hired trackers to warn the Mukogodo herders of pending baboon attacks on young goats.

All parties benefited from Shirley's conservation-cum-development work until the Mukogodo herders switched from mobile herding to permanent settlement in the late 1980s and degraded the land. Bushlands invaded the denuded grasslands, making life harder for herders on the heavily eroded and gullied landscape. The baboons did just fine on the flowers and fruits of the invading thickets until the herders began cutting down the trees for charcoal to make ends meet. Colonizing the overgrazed land, a cactus from the southeastern United States, *Opuntia stricta*, spread prolifically, causing consternation among biologists worried about an alien invasive species disrupting the ecology. Whatever the provenance of the cactus, the fecund opuntia fruits not only boosted baboon growth and reproductive rates to those of the Pumphouse Gang in its crop-raiding days on Kekopey, but also filled a void in the diet of elephants suffering a loss of woody vegetation to charcoal makers during the 2009 drought.

The baboons' flexible behavioral response to every change in human activity at Chololo continues, pointing to the need for an adaptable pluralistic approach to land and species conservation.

Property and resource rights are not equal among humans themselves, let alone humans interacting with other species. Self-interest, ethnicity, and dominance come into play. Only in the global age have human rights been universally adopted, in the UN Charter for Human Rights of 1948. Even now women, children, ethnic minorities, and the poor are still denied fundamental human rights over much of the world.

When it comes to other species, some, as in George Orwell's satirical view of communism in *Animal Farm,* enjoy more rights than others, depending on whether they are pleasing, appealing, troubling, or dangerous to us. What if the animals struggling to find a place on Kekopey were not baboons with Shirley to champion them, but rodents or weevils eating the stored grain of farmers, or cockroaches invading their homes? Kenyans view species on the same sliding scale of useful, pestilent, and appealing as Americans who spend billions of dollars eradicating pest species and millions more saving the wolf, bear, and bison.

We grant animals rights only once we have won our own rights, and then only selectively. Belief in the intrinsic rights of species assumes a consciousness of rights in other animals. Yet aside from us, species don't grant rights to other species any more than they recognize and respect our own property rights. Animals have even less power to protect their rights than hunter-gatherers did theirs when challenged by pastoralists, and pastoralists did theirs when evicted by the British colonial government.

Would baboons, or any other species achieving superdominance, grant humans any property rights or set aside national parks to conserve us? Baboons have neither the cooperative skills nor the concern for other species to do so. Evidence is mounting of altruism among other species, such as dolphins saving humans from drowning, perhaps, but no other species has escaped Charles Lyell's "economy of nature." We alone intentionally protect other species, regardless of their utility.

The clear distinction between natural and humanized habitats, to which baboons are oblivious, becomes a matter of degree anyway as we reengineer landscapes and ecosystems. All ecosystems are now shaped directly or indirectly by human activity in varying degrees. In Africa, our ancestral home, we have been modifying the ecology for half a million years, and in the Americas fewer than fifteen thousand years, though far more extensively.

Baboons, like coyotes, bears, and raccoons, don't distinguish between natural and human foods either, or between indigenous and exotic species. They use whatever foods appear best and easiest to harvest. In changing their range and adapting their behavior, baboons reflect the new norm for countless species adapting to human-dominated landscapes and selective pressures. Shirley started out with a passion for research before turning conservationist to save the baboons she studies. Before long she was helping the community suffering baboon depredations to deflect competition and make baboons an asset.

We must accept the complicated intertwining of nature and humanity in what is after all the Human Age and adapt conservation solutions to local conditions. I believe we can make headway finding space for wildlife in the humanized landscapes of today rather than relying solely on protected areas. Conservation in the global age should recognize the welter of rights, values, and views of nature and figure out how to reconcile the hodgepodge of interests within the democratic process.

Joshua Greene, in *Moral Tribes*, argues that philosophers have tried and failed for millennia to find self-evident foundations on which to build universal principles and extend moral consideration to others beyond our own family and tribe.[5] Moral philosophers and theologians have looked for abstract truths in the human mind rather than in the messy human nature emerging out of our primate ancestry and modified in the course of our cultural evolution. Our brain is geared to respond to proximate, visible, and tangible perceptions rather than the remote impersonal challenges of globalization.

Our built-in emotional responses falter when we scale up our activities over the horizon and deal with alien peoples and cultures. Ethnocentricity is a universal. Overcoming in-group glue calls for a meta-morality, according to Greene—the ability to bridge tribal differences in beliefs: views of nature, religion, politics, fairness, and the freedom of the individual relative to group responsibilities. As I see it, because other species are not just another person but whole other beings alien in varying degrees, depending on how we relate to them, conserving biodiversity calls for the highest level of consideration, rationalization, and altruism.

Peter Singer, in *The Expanding Circle*, argues that extending our sense of fairness and consideration from those close by and familiar to other people and other species calls for a cognitive leap.[6] I have argued instead that our expanding sensibilities and concern for other species emerged not from a cognitive leap or novel genes but from our ecological emancipation, cultural enlightenment, and modern sensibilities. The best hope I see for expanding our consideration to other species lies in speeding up the demographic and economic transition worldwide so that all of us can afford and enjoy the wealth of our cultural and biological heritage.

Ethics philosopher Bryan Norton calls for a pluralistic approach to conservation, rooted in the principle of treating others as you would yourself and extending similar consideration to future generations.[7] Conserving biodiversity and sustaining planetary health require planning on a geographical and evolutionary scale far beyond our emotional, rational, and national framework. With differences between tribes, cultures, religions, and rich and poor nations still so wide, the focus should be on bridging the gaps in opportunities. A unity in aspirations will grow out of narrowing the divides and bettering well-being among all nations. We have made a good deal of progress in forging international goals and agreements over the last half century. If each nation cleans its own nest and doesn't impose costs on others, the larger common goals can be met. If not, we are headed for a global tragedy of the commons from which there is no escape hatch.

I value wildlife for its own sake and wish for others to feel the same way. I can't bank on most people feeling the same way anytime soon, though. In the meantime, the best prospects lie in promoting a diversity of conservation goals, whether for water, food security, energy, environmental health, well-being, biodiversity, or planetary health. Saving natural heritage such as the Serengeti or cultural heritage such as *The Wedding at Cana* comes down to those things we care for. The extraordinary reach and depth of conservation beggar any unitary philosophy, policy, and method. That we are a rambunctious species by nature shouldn't lull us into the fatalism of thinking we are inherently destructive. We should take heart that of all species that might have risen to global dominance, we alone have the sentience, knowledge, foresight, and capacity to make the planet a fit place for our descendants and all life.[8]

Notes

1. Survival in the Savannas

1. Joseph Thomson, *Through Maasai Land* (London: S. Low, Marston, Searle, & Rivington, 1885).

2. David Western, *In the Dust of Kilimanjaro* (Washington, DC: Island, 2001).

3. David Western and Virginia Finch, "Cattle and Pastoralism: Survival and Production in Arid Lands," *Human Ecology* 14, no. 1 (1986): 77–94.

4. Alfred Claud Hollis, *The Masai: Their Language and Folklore* (Oxford: Clarendon, 1905); Katherine M. Homewood and Allan W. Rodgers, *Maasailand Ecology: Pastoralist Development and Wildlife Conservation in Ngorongoro, Tanzania* (Cambridge: Cambridge University Press, 1991); Naomi Kipury, *Oral Literature of the Maasai* (Portsmouth, NH: Heinemann Educational Books, 1983); Tepilit Ole Saitoti and Carol Beckwith, *Maasai* (New York: Abrams, 1980).

5. Virginia Finch and David Western, "Cattle Colors in Pastoral Herds: Natural Selection or Social Preference?" *Ecology* 58, no. 6 (1977): 1384–92.

6. David Western, "Ecosystem Conservation and Rural Development: The Case of Amboseli," in *Natural Connections: Perspectives in Community-Based Conservation*, ed. David Western, R. Michael Wright, and Shirley C. Strum (Washington, DC: Island, 1994).

7. The following sources provide further background reading: Jonathan S. Adams and Thomas O. McShane, *The Myth of Wild Africa: Conservation without Illusion* (Berkeley: University of California Press, 1996); David Anderson and Richard H. Grove, *Conservation in Africa: Peoples, Policies and Practice* (Cambridge: Cambridge University Press, 1989); Galaty G. John and Johnson L. Douglas, eds., *The World of Pastoralism* (New York: Guilford, 1990); David Western, John Waithaka, and John Kamanga, "Finding Space for Wildlife beyond National Parks and Reducing Conflict through Community-Based Conservation: The Kenya Experience," *Parks* 21, no. 1 (2015): 51–62; Roderick P. Neumann, "Africa's Last Wilderness: Reordering Space for Political

and Economic Control in Colonial Tanzania," *Journal of the International African Institute* 71, no. 4 (2001): 641–65; Tom Bourner, *Homo Prospectus, Action Learning: Research and Practice* (Oxford: Oxford University Press, 2017), https://doi.org/10.1080/14767333.2017.1315218; David Western, "The Environment and Ecology of Pastoralists in Arid Savannas," *Development and Change* 13, no. 2 (1982): 183–211.

2. Consuming Passions

1. Martha Honey, *Ecotourism and Sustainable Development: Who Owns Paradise?* (Washington, DC: Island, 2008).

2. Mark Dowie, "The Hidden Cost of Paradise," *Stanford Social Innovation Review* 4, no. 1 (2006): 28–36.

3. Raymond Bonner, *At the Hand of Man: Peril and Hope for Africa's Wildlife* (New York: Vintage, 1993); Marc Chapin, "A Challenge to Conservationists," *World Watch*, November–December, 2004.

4. U.S. Environmental Protection Agency, *Our Nation's Air: Status and Trends through 2015*, 2016, https://gispub.epa.gov/air/trendsreport/2016/.

5. The following sources provide further background reading: U.S. Environmental Protection Agency, *Our Nation's Air*; Daudi Peterson, Richard Baalow, and Jon Cox, *Hadzabe: By the Light of a Million Fires* (Dar es Salaam: Mkuki na Nyota, 2013); Roderick P. Neumann, *Imposing Wilderness: Struggles over Livelihood and Nature Preservation in Africa* (Berkeley: University of California Press, 1998); Abraham Harold Maslow, *Motivation and Personality* (New York: Harper & Row, 1954).

3. The Conservation Paradox

1. Pasquale Gagliardi, Bruno Latour, and Pedro Memelsdorff, *Coping with the Past: Creative Perspectives on Conservation and Restoration* (Florence: Casa Editrice Leo S. Olschki, 2010).

2. R. K. Pachauri and L. A. Meyer, eds., *Climate Change 2014: Synthesis Report; Contribution of Working Groups I, II and III to the Fifth Assessment Report of the Intergovernmental Panel on Climate Change* (Geneva: IPCC, 2015).

3. Charles Darwin, *The Voyage of the* Beagle (London: Penguin, 1989).

4. Charles L. Redman, *Human Impact on Ancient Environments* (Tucson: University of Arizona Press, 1999).

5. Abraham Harold Maslow, *Motivation and Personality* (New York: Harper & Row, 1954).

6. The following source provides further background reading: Peter Coates, *Nature: Western Attitudes since Ancient Times* (Cambridge, MA: Polity, 1998).

4. Limits to Growth

1. Paul Ehrlich, *The Population Bomb* (New York: Ballantine Books, 1968).

2. Julian Lincoln Simon and Herman Kahn, eds., *The Resourceful Earth: A Response to Global 2000* (New York: Blackwell, 1984).

3. Mary Tiffen, Michael Mortimore, and Francis Gichuki, *More People, Less Erosion: Environmental Recovery in Kenya* (New York: John Wiley & Sons, 1994).

4. Ester Boserup, *Conditions of Agricultural Growth: The Economics of Agrarian Change under Population Pressure* (New York: Routledge, 1965).

5. Thomas Malthus, *An Essay on the Principle of Population, as It Affects the Future Improvement of Society* (New York: Cosimo, 2013).

6. Adam Smith, *The Wealth of Nations* (New York: Bantam Dell, 2003).

7. Donella H. Meadows et al., *The Limits to Growth: A Report for the Club of Rome's Project on the Predicament of Mankind* (New York: Universe Books, 1974).

8. Rachel Carson, *Silent Spring* (Boston: Houghton Mifflin, 1962).

9. Barry Commoner, *The Closing Circle: Nature, Man, and Technology* (New York: Bantam, 1968).

10. Matt Ridley, *The Rational Optimist: How Prosperity Evolves* (New York: Harper, 2012).

11. Bjørn Lomborg, *The Skeptical Environmentalist: Measuring the Real State of the World*, vol. 1 (Cambridge: Cambridge University Press, 2001).

12. David I. Stern, "Progress on the Environmental Kuznets Curve?" *Environment and Development Economics* 3, no. 2 (1998): 173–96.

13. Thomas Picketty, *Capital in the Twenty-First Century* (Cambridge, MA: Harvard University Press, 2014).

14. Ridley, *The Rational Optimist*; Lomborg, *The Skeptical Environmentalist*.

15. David Western and Mary C. Pearl, *Conservation for the Twenty-First Century* (New York: Oxford University Press, 1989).

16. Fairfield Osborn, *Our Plundered Planet* (Boston: Little, Brown, 1948).

17. Gro Harlem Brundtland, *Our Common Future: The Report of the World Commission on Environment and Development* (New York: United Nations, 1987).

18. Jared Diamond, *Collapse: How Societies Choose to Fail or Succeed* (New York: Penguin, 2005); Jared Diamond, *Guns, Germs, and Steel: The Fate of Human Societies* (New York: Norton, 1997).

19. Ryan D. Bergstrom, *Questioning Collapse: Human Resilience, Ecological Vulnerability, and the Aftermath of Empire*, ed. Patricia A. MacAnany and Norman Yoffee (Cambridge: Cambridge University Press, 2010).

20. Picketty, *Capital in the Twenty-First Century*.

21. Bergstrom, *Questioning Collapse*.

22. Nick Brooks et al., "The Archaeology of Western Sahara: Results of Environmental and Archaeological Reconnaissance," *Antiquity* 83, no. 322 (2009): 918–34.

23. The following sources provide further background reading: Bill McKibben, *The End of Nature* (New York: Random House, 2006); Tim Flannery, *Here on Earth: A Twin Biography of the Planet and the Human Race* (London: Penguin, 2012); National Research Council, *Population Growth and Economic Development: Policy Questions* (Washington, DC: National Academies Press, 1986); William Cronon, ed., *Uncommon Ground: Rethinking the Human Place in Nature* (New York: Norton, 1996); Global 2000 Study (U.S.), *The Global 2000 Report to the President—Entering the Twenty-First Century*, vol. 1 (Ann Arbor: University of Michigan Library, 1980).

5. Lessons from Disasters

1. Gregg Easterbrook, *The Progress Paradox: How Life Gets Better While People Feel Worse* (New York: Random House, 2003).

2. Arthur Herman, *The Idea of Decline in Western History* (New York: Free Press, 1997).

3. Allan William, *The African Husbandman* (Edinburgh: Oliver & Boyd, 1965).

4. Susan L. Cutter et al., "Global Risks: Pool Knowledge to Stem Losses from Disasters," *Nature News* 522, no. 7556 (2015): 277.

5. Henrik Svensen, *The End Is Nigh: A History of Natural Disasters* (London: Reaktion Books, 2009).

6. Kenneth Hewitt, *Regions of Risk: A Geographical Introduction to Disasters* (London: Routledge, 2014).

7. David Western et al., "Predicting Extreme Droughts in Savannah Africa: A Comparison of Proxy and Direct Measures in Detecting Biomass Fluctuations, Trends and Their Causes," *PLoS* 10, no. 8 (2015): e0136516.

8. Jason Burke, "Muhammad Yunus Appeals to the West to Help Bangladesh's Garment Industry," *Guardian*, May 12, 2013.

9. The following sources provide further background reading: David R. Montgomery, *Dirt: The Erosion of Civilizations* (Berkeley: University of California Press, 2007); Keith Smith, *Environmental Hazards: Assessing Risk and Reducing Disaster* (London: Routledge, 2013); Kathleen Tierney, *The Social Roots of Risk: Producing Disasters, Promoting Resilience* (Stanford, CA: Stanford University Press, 2014).

6. Why Some Succeed Where Others Fail

1. Garrett Hardin, "The Tragedy of the Commons," *Science* 162, no. 3859 (1968): 1243–48.

2. Partha Dasgupta, *Human Well-Being and the Natural Environment* (Oxford: Oxford University Press, 2001).

3. Paul Schultz Martin, *Twilight of the Mammoths: Ice Age Extinctions and the Rewilding of America* (Berkeley: University of California Press, 2005).

4. Ross D. E. MacPhee, *Extinctions in Near Time: Causes, Context, and Consequences* (New York: Plenum, 2003).

5. Cited in Elinor Ostrom, *Governing the Commons: The Evolution of Institutions for Collective Action* (Cambridge: Cambridge University Press, 1990).

6. David Western, "The Environment and Ecology of Pastoralists in Arid Savannas," *Development and Change* 13, no. 2 (1982): 183–211; Brian Spooner and Haracharan Singh Mann, eds., *Desertification and Development: Dryland Ecology in Social Perspective* (London: Academic, 1982); S. Sandford, "Pastoral Strategies and Desertification: Opportunism and Conservatism in Dry Lands," in Spooner and Mann, *Desertification and Development*. Quote: Garrett Hardin, "Extensions of 'The Tragedy of the Commons,'" *Science* 280 (1998): 682–83.

7. David Western, "The Background to Community-Based Conservation," in *Natural Connections: Perspectives in Community-Based Conservation*, ed. David

Western, R. Michael Wright, and Shirley C. Strum (Washington, DC: Island, 1994), 1–12.

8. Ostrom, *Governing the Commons.*

9. Susan J. Buck, *The Global Commons: An Introduction* (Washington, DC: Island, 1998).

10. Joshua E. Cinner et al., "Bright Spots among the World's Coral Reefs," *Nature* 535, no. 7612 (2016): 416.

11. Robert B. Reich, *The Common Good* (New York: Vintage, 2019).

12. The following sources provide further background reading: Charles R. Lane, ed., *Custodians of the Commons: Pastoral Land Tenure in East and West Africa* (London: Earthscan, 1998); Fikret Berkes, Carl Folke, and Johan Colding, eds., *Linking Social and Ecological Systems: Management Practices and Social Mechanisms for Building Resilience* (Cambridge: Cambridge University Press, 2000); Peter Turchin, *Ultra-Sociality: How 10,000 Years of War Made Humans the Greatest Cooperators on Earth* (Chaplin, CT: Beresta Books, 2015).

7. Icons of Two Worlds

1. Alfred E. Kahn, "The Tyranny of Small Decisions: Market Failures, Imperfections, and the Limits of Economics," *Kyklos* 19, no. 1 (1966): 23–47; William E. Odum, "Environmental Degradation and the Tyranny of Small Decisions," *BioScience* 32, no. 9 (1982): 728–29.

2. Nathan Sayre, *Working Wilderness: The Malpai Borderlands Group Story and the Future of the Western Range* (Tucson: Rio Nuevo, 2006).

3. Wallace Stegner, *Beyond the Hundredth Meridian: John Wesley Powell and the Second Opening of the West* (New York: Penguin, 1992).

4. Courtney White, *Revolution on the Range: The Rise of a New Ranch in the American West* (Washington, DC: Island, 2012).

5. Charles Curtin and David Western, "Grasslands, People, and Conservation: Over-the-Horizon Learning Exchanges between African and American Pastoralists," *Conservation Biology* 22, no. 4 (2008): 870–77.

6. Paul Schultz Martin, *Twilight of the Mammoths: Ice Age Extinctions and the Rewilding of America* (Berkeley: University of California Press, 2005).

7. Warner Glenn, *Eyes of Fire: Encounter with a Borderlands Jaguar* (New York: Treasure Chest, 1996).

8. Liz Rosan, *Preserving Working Ranches in the West* (Tucson: Sonoran Institute, 1997).

9. Margaret Mead, quoted in Donald Keys, *Earth Omega: Passage to Planetization* (Boston: Brandon, 1982).

10. The following sources provide further background reading: John Baden and Donald Snow, eds., *The Next West: Public Lands, Community, and Economy in the American West* (Washington, DC: Island, 1997); Charles G. Curtin, *The Science of Open Spaces: Theory and Practice for Conserving Large, Complex Systems* (Washington, DC: Island, 2015); Dana L. Jackson and Laura L. Jackson, eds., *The Farm as Natural Habitat:*

Reconnecting Food Systems with Ecosystems (Washington, DC: Island, 2002); Daniel Kemmis, *This Sovereign Land: A New Vision for Governing the West* (Washington, DC: Island, 2001); James P. Owen, *Cowboy Values: Recapturing What America Once Stood For* (Guilford, CT: Lyons, 2008); Thomas Michael Power and Richard Barrett, *Post-Cowboy Economics: Pay and Prosperity in the New American West* (Washington, DC: Island, 2001).

8. An Altruistic Species

1. Charles Lyell, *Principles of Geology,* vol. 3 (London, 1833).

2. Charles Darwin, *The Descent of Man and Selection in Relation to Sex* (London: John Murray, 1871).

3. Charles Darwin, *On the Origin of Species: By Means of Natural Selection* (London: John Murray, 1859).

4. Mark Kurlansky, *Cod: A Biography of the Fish That Changed the World* (New York: Penguin, 1997).

5. Alfred Russel Wallace, *The Geographical Distribution of Animals with a Study of the Relations of Living and Extinct Fauna as Elucidating the Past Changes of the Earth's Surface,* vol. 1 (New York: Harper & Brothers, 1876); Alfred Russel Wallace, *The World of Life: A Manifestation of Creative Power, Directive Mind and Ultimate Purpose* (New York: Moffat, Yard, 1911).

6. Edward O. Wilson, *Biophilia: The Human Bond with Other Species* (Cambridge, MA: Harvard University Press, 1984).

7. David Sloan Wilson, *Does Altruism Exist? Culture, Genes, and the Welfare of Others* (New Haven, CT: Yale University Press, 2015).

8. W. Hamilton, "The Genetical Evolution of Social Behavior," *Journal of Theoretical Biology* 7, no. 1 (1964): 1–6; Edward O. Wilson, *Sociobiology: The New Synthesis* (Cambridge, MA: Harvard University Press, 1975).

9. Richard Dawkins, *The Selfish Gene* (Oxford: Oxford University Press, 1976).

10. Ernst Fehr and Simon Gächter, "Altruistic Punishment in Humans," *Nature* 10 (2002): 137–40.

11. Richard Dawkins, *A Devil's Chaplain: Reflections on Hope, Lies, Science, and Love* (Boston: Mariner Books, 2003).

12. Richard Dawkins, "An Open Letter to Prince Charles," *Edge,* 2000, https://www.edge.org/3rd culture/prince/prince index. html.

13. Wilson, *Biophilia*.

14. Stephen R. Kellert, "The Biological Basis for Human Values of Nature," and Jared Diamond, "New Guineans and Their Natural World," both in *The Biophilia Hypothesis,* ed. Stephen R. Kellert and Edward O. Wilson (Washington, DC: Island, 1993).

15. Guillaume Chomicki and Susanne S. Renner, "Obligate Plant Farming by a Specialized Ant," *Nature Plants* 2, no. 12 (2016): 16181.

16. Edward O. Wilson, *The Social Conquest of Earth* (New York: Liveright, 2013).

9. Breaking Biological Barriers

1. Shirley C. Strum, *Almost Human: A Journey into the World of Baboons* (New York: Random House, 1987).

2. Charles Darwin, *The Descent of Man and Selection in Relation to Sex* (London: John Murray, 1871).

3. Nicholas K. Humphrey, "The Social Function of Intellect," in *Growing Points in Ethology,* ed. P. P. G. Bateson and R. A. Hinde (Cambridge: Cambridge University Press, 1976), 303–17.

4. Richard Byrne and Andrew Whiten, eds., *Machiavellian Intelligence: Social Expertise and the Evolution of Intellect in Monkeys, Apes, and Humans* (Oxford: Clarendon, 1988).

5. Adam Smith, *The Wealth of Nations* (New York: Bantam Dell, 2003).

6. Lawton Graham, ed., *The Human Story* (New York: Reed Business Information, 2014).

7. Michael Balter, "The Killing Ground," *Science* 344, no. 6188 (2014): 180–83.

8. Felisa A. Smith et al., "Body Size Downgrading of Mammals over the Late Quaternary," *Science* 360, no. 6386 (2018): 310–13.

9. Daniel L. Everett, *How Language Began: The Story of Humanity's Greatest Invention* (New York: Liveright, 2017).

10. Robin I. M. Dunbar, "Neocortex Size as a Constraint on Group Size in Primates," *Journal of Human Evolution* 22, no. 6 (1992): 469–93.

11. Jonah D. Western and Shirley C. Strum, "Sex, Kinship, and the Evolution of Social Manipulation," *Ethology and Sociobiology* 4, no. 1 (1983): 19–28.

12. Michael Tomasello, *The Cultural Origins of Human Cognition* (Cambridge, MA: Harvard University Press, 1999).

13. Frans de Waal, *The Age of Empathy: Nature's Lessons for a Kinder Society* (New York: Crown, 2010).

14. Martin E. P. Seligman et al., *Homo Prospectus* (Oxford: Oxford University Press, 2016).

15. Marcus J. Hamilton et al., "The Complex Structure of Hunter-Gatherer Social Networks," *Proceedings of the Royal Society B: Biological Sciences* 274, no. 1622 (2007): 2195–2203.

16. D'Arcy Wentworth Thompson, *On Growth and Form* (Cambridge: Facsimile, 1917).

17. Max Kleiber, *The Fire of Life: An Introduction to Animal Energetics* (New York: John Wiley & Sons, 1961).

18. David Western, "Size, Life History and Ecology in Mammals," *African Journal of Ecology* 17, no. 4 (1979): 185–204.

19. Geoffrey B. West, *Scale: The Universal Laws of Growth, Innovation, Sustainability, and the Pace of Life in Organisms, Cities, Economies, and Companies* (New York: Penguin, 2017).

20. Leslie C. Aiello and Peter Wheeler, "The Expensive-Tissue Hypothesis: The Brain and the Digestive System in Human and Primate Evolution," *Current Anthropology* 36, no. 2 (1995): 199–221.

21. Richard Wrangham, *Catching Fire: How Cooking Made Us Human* (New York: Basic Books, 2009).

22. Katherine D. Zink and Daniel E. Lieberman, "Impact of Meat and Lower Palaeolithic Food Processing Techniques on Chewing in Humans," *Nature* 531, no. 7595 (2016): 500.

23. Herman Pontzer et al., "Metabolic Acceleration and the Evolution of Human Brain Size and Life History," *Nature* 533, no. 7603 (2016): 390.

24. Peter Turchin, *War and Peace and War: The Life Cycles of Imperial Nations* (New York: Pi, 2005).

25. Elinor Ostrom, *Governing the Commons: The Evolution of Institutions for Collective Action* (Cambridge: Cambridge University Press, 1990).

26. Joseph Henrich et al., "Costly Punishment across Human Societies," *Science* 312 (2006): 1767–70.

27. Adam Smith, *The Theory of Moral Sentiments* (Edinburgh, 1759).

28. David Wilson, *Darwin's Cathedral: Evolution, Religion, and the Nature of Society* (Chicago: University of Chicago Press, 2003).

29. Ernst Fehr and Simon Gächter, "Altruistic Punishment in Humans," *Nature* 10 (2002): 137–40.

30. Joshua David Greene, *Moral Tribes: Emotion, Reason, and the Gap between Us and Them* (New York: Penguin, 2013).

31. Frans de Waal, *Are We Smart Enough to Know How Smart Animals Are?* (New York: Norton, 2016).

32. Darwin, *The Descent of Man*.

33. David Sloan Wilson, *Does Altruism Exist? Culture, Genes, and the Welfare of Others* (New Haven, CT: Yale University Press, 2016).

34. The following sources provide further background reading: Robert Wright, *Nonzero: History, Evolution & Human Cooperation* (London: Abacus, 2001); Yuval Noah Harari, *Sapiens: A Brief History of Humankind* (New York: Harper, 2015); Robert Wright, *The Moral Animal: Evolutionary Psychology and Everyday Life* (New York: Vintage Books, 1995); Edward O. Wilson, *The Social Conquest of Earth* (New York: Liveright, 2013); Itai Yanai and Lercher Martin, *The Society of Genes* (Cambridge, MA: Harvard University Press, 2016).

10. Domesticating Nature

1. David Reich, *Who We Are and How We Got Here: Ancient DNA and the New Science of the Human Past* (New York: Vintage, 2018).

2. Vaclav Smil, *The Earth's Biosphere: Evolution, Dynamics, and Change* (Cambridge, MA: MIT Press, 2003); Thomas Suddendorf, *The Gap: The Science of What Separates Us from Other Animals* (New York: Basic Books, 2013).

3. Mark Lynas, *The God Species: How Humans Really Can Save the Planet* . . . (London: Fourth Estate, 2012).

4. L. M. Talbot and M. H. Talbot, "The Biological Productivity of Tropical Savanna Ecosystems," in *The Ecology of Man in the Tropical Environment,* IUCN new series, no. 4 (Morges: IUCN, 1964), 88–97; P. Arman and D. Hopcraft, "Nutritional Studies of East African Herbivores, I: Digestibility of Dry Matter, Crude Fiber and Crude Protein in Antelope, Cattle and Sheep," *British Journal of Nutrition* 33, no. 2 (1975): 255–64.

5. Andrew N. Muchiru, David Western, and Robin S. Reid, "The Impact of Abandoned Pastoral Settlements on Plant and Nutrient Succession in an African Savanna Ecosystem," *Journal of Arid Environments* 73, no. 3 (2009): 322–31.

6. Leah H. Samberg, Carol Shennan, and Erika S. Zavaleta, "Human and Environmental Factors Affect Patterns of Crop Diversity in an Ethiopian Highland Agroecosystem," *Professional Geographer* 62, no. 3 (2010): 395–408.

7. Richard Phillips Feynman, *The Meaning of It All: Thoughts of a Citizen Scientist* (New York: Perseus Books, 1998).

8. Ester Boserup, *The Conditions of Agricultural Growth: The Economics of Agrarian Change under Population Pressure* (London: Earthscan, 1965).

9. James C. Scott, *Against the Grain: A Deep History of the Earliest States* (New Haven, CT: Yale University Press, 2017).

10. Monica N. Ramsey and Arlene M. Rosen, "Wedded to Wetlands: Exploring Late Pleistocene Plant-Use in the Eastern Levant," *Quaternary International* 396 (2016): 5–19.

11. Richard C. Francis, *Domesticated: Evolution in a Man-Made World* (New York: Norton, 2015).

12. Simone Riehl, Mohsen Zeidi, and Nicholas J. Conard, "Emergence of Agriculture in the Foothills of the Zagros Mountains of Iran," *Science* 341, no. 6141 (2013): 65–67.

13. Francis, *Domesticated.*

14. Charles Darwin, *The Variation of Animals and Plants under Domestication,* vol. 2 (New York: D. Appleton, 1894).

15. Tecumseh Fitch, "How Pets Got Their Spots (and Floppy Ears)," *New Scientist* 225, no. 3002 (2015): 24–25.

16. Gary Paul Nabhan, *Where Our Food Comes From: Retracing Nikolay Vavilov's Quest to End Famine* (Washington, DC: Island, 2009).

17. Adam Smith, *The Wealth of Nations* (New York: Bantam Dell, 2003).

18. Peter J. Richerson and Robert Boyd, *Not by Genes Alone: How Culture Transformed Human Evolution* (Chicago: University of Chicago Press, 2005).

19. Joseph Henrich, *The Secret of Our Success: How Culture Is Driving Human Evolution, Domesticating Our Species, and Making Us Smarter* (Princeton, NJ: Princeton University Press, 2017).

20. Thomas Talhelm et al., "Large-Scale Psychological Differences within China Explained by Rice versus Wheat Agriculture," *Science* 344, no. 6184 (2014): 603–8.

21. Henrich, *The Secret of Our Success.*

11. Ecological Emancipation

1. Ibn Khaldun, *The Muqaddimah: An Introduction to History* (Princeton, NJ: Princeton University Press, 1967).

2. Robert T. Paine, "A Note on Trophic Complexity and Community Stability," *American Naturalist* 103, no. 929 (1969): 91–93.

3. Nelson G. Hairston, Frederick E. Smith, and Lawrence B. Slobodkin, "Community Structure, Population Control, and Competition," *American Naturalist* 94, no. 879 (1960): 421–25.

4. Thomas Picketty, *Capital in the Twenty-First Century* (Cambridge, MA: Harvard University Press, 2014).

5. Adam Smith, *The Wealth of Nations* (New York: Bantam Dell, 2003).

6. Picketty, *Capital in the Twenty-First Century.*

7. Bruce L. Gardner, *American Agriculture in the Twentieth Century: How It Flourished and What It Cost* (Cambridge, MA: Harvard University Press, 2009).

8. Robert J. Gordon, *The Rise and Fall of American Growth: The U.S. Standard of Living since the Civil War* (Princeton, NJ: Princeton University Press, 2017).

9. David Western, "Human-Modified Ecosystems and Future Evolution," *Proceedings of the National Academy of Sciences* 98, no. 10 (2001): 5458–65.

10. Paul Roberts, *The End of Food: The Coming Crisis in the World Food Industry* (Boston: Houghton Mifflin, 2008); Bill McKibben, *Eaarth: Making a Life on a Tough New Planet* (New York: St. Martin's Griffin, 2011).

11. Francis Fukuyama, *The Origins of Political Order: From Prehuman Times to the French Revolution* (New York: Farrar, Straus & Giroux, 2011).

12. Ronald Wright, *A Short History of Progress* (Toronto: House of Anansi, 2004).

13. Joseph A. Tainter, *The Collapse of Complex Societies* (Cambridge: Cambridge University Press, 1988).

14. Atsushi Iriki and Osamu Sakura, "The Neuroscience of Primate Intellectual Evolution: Natural Selection and Passive and Intentional Niche Construction," *Royal Society London* 363, no. 1500 (2008): 2229–41; Timothy Taylor, *The Artificial Ape: How Technology Changed the Course of Human Evolution* (New York: St. Martin's, 2010).

15. Jared Diamond, *Guns, Germs, and Steel: The Fate of Human Societies* (New York: Norton, 1997).

16. Fukuyama, *The Origins of Political Order.*

12. The Global Express

1. Tavneet Suri and William Jack, "The Long-Run Poverty and Gender Impacts of Mobile Money," *Science* 354, no. 6317 (2016): 1288–92.

2. Thomas P. Hughes, *Human-Built World: How to Think about Technology and Culture* (Chicago: University of Chicago Press, 2004).

3. Manfred B. Steger, *Globalization: A Very Short Introduction* (Oxford: Oxford University Press, 2017). For other views of globalization, see Jan Aart Scholte, *Globalization: A Critical Introduction* (Melbourne: Macmillan International Higher Educa-

tion, 2005); Patricia J. Campbell, Aran MacKinnon, and Christy R. Stevens, *An Introduction to Global Studies* (Hoboken, NJ: Wiley-Blackwell, 2010).

4. David Ricardo, *On the Principles of Political Economy and Taxation* (London: John Murray, 1817).

5. Donald R. Davis, "Intra-industry Trade: A Heckscher-Ohlin-Ricardo Approach," *International Economics* 39, no. 3–4 (1995): 201–26.

6. Joseph E. Stiglitz, *Globalization and Its Discontents* (New York: Norton, 2002).

7. Robert Gilpin, *The Challenge of Global Capitalism: The World Economy in the 21st Century* (Princeton, NJ: Princeton University Press, 2002).

8. Thomas L. Friedman, *The World Is Flat: A Brief History of the Twenty-First Century* (New York: Farrar, Straus & Giroux, 2005).

9. Andre Gunder Frank, *World Accumulation* (London: Macmillan, 1978).

10. Alison S. Brook et al., "Long-Distance Stone Transport and Pigment Use in the Earliest Middle Stone Age," *Science* 360, no. 6384 (2018): 90–94.

11. Barry Wellman and Lee Rainie, *Networked: The New Social Operating System* (Cambridge, MA: MIT Press, 2012); Mark Peplow, "Organic Synthesis: The Robo-Chemist," *Nature News* 512, no. 7512 (2014): 20.

13. Converging Worlds

1. Francesco D'Errico et al., "Early Evidence of San Material Culture Represented by Organic Artifacts from Border Cave, South Africa," *Proceedings of the National Academy of Sciences* 109, no. 33 (2012): 13214–19.

2. Thomas L. Friedman, *The Lexus and the Olive Tree: Understanding Globalization* (New York: Farrar, Straus & Giroux, 1999).

3. William W. Lewis, *The Power of Productivity: Wealth, Poverty, and the Threat to Global Stability* (Chicago: University of Chicago Press, 2005).

4. Thomas Picketty, *Capital in the Twenty-First Century* (Cambridge, MA: Harvard University Press, 2014).

5. Timothy Hall Breen, *The Marketplace of Revolution: How Consumer Politics Shaped American Independence* (New York: Oxford University Press, 2004).

6. Manfred B. Steger, *Globalization: A Very Short Introduction* (Oxford: Oxford University Press, 2017).

7. Joseph E. Stiglitz, *Globalization and Its Discontents* (New York: Norton, 2002); Will Hutton, *A Declaration of Interdependence: Why America Should Join the World* (New York: Norton, 2004).

8. Robert Gilpin, *The Challenge of Global Capitalism: The World Economy in the 21st Century* (Princeton, NJ: Princeton University Press, 2002).

9. Edward Alden, *Failure to Adjust: How Americans Got Left Behind in the Global Economy* (Lanham, MD: Rowman & Littlefield, 2017).

10. Joseph S. Nye, *The Future of Power* (New York: Public Affairs, 2011).

11. Howard Rheingold, *Smart Mobs: The Next Social Revolution* (New York: Basic Books, 2003).

12. Elinor Ostrom, *Governing the Commons: The Evolution of Institutions for Collective Action* (Cambridge: Cambridge University Press, 1990).

13. Robert B. Reich, *Supercapitalism: The Transformation of Business, Democracy, and Everyday Life* (New York: Vintage Books, 2008).

14. John Dewey, *The Public and Its Problems* (New York: Henry Holt, 1927).

15. Gross Neil, "Is the United States Too Big to Govern?" *New York Times,* May 11, 2018.

16. Robert D. Putnam, *Bowling Alone: The Collapse and Revival of American Community* (New York: Touchstone Books, 2001).

17. James Fallows and Deborah Fallows, *Our Towns: A 100,000-Mile Journey into the Heart of America* (New York: Pantheon Books, 2018).

18. Kevin Kelly, *The Inevitable: Understanding the 12 Technological Forces That Will Shape Our Future* (New York: Penguin, 2017).

14. Our Novel Age

1. Marlene Zuk, *Paleofantasy: What Evolution Really Tells Us about Sex, Diet, and How We Live* (New York: Norton, 2013).

2. Michael J. Sandel, *Justice: What's the Right Thing to Do?* (New York: Farrar, Straus & Giroux, 2010).

3. Manfred B. Steger, *Globalization: A Very Short Introduction* (Oxford: Oxford University Press, 2017).

4. Joshua David Greene, *Moral Tribes: Emotion, Reason, and the Gap between Us and Them* (New York: Penguin, 2013).

5. James Suzman, *Affluence without Abundance: What We Can Learn from the World's Most Successful Civilisation* (New York: Bloomsbury, 2017); Michael J. Sandel, *What Money Can't Buy: The Moral Limits of Markets* (New York: Farrar, Straus & Giroux, 2013).

6. Paul J. Crutzen, "Geology of Mankind," *Nature* 415 (2002): 23.

7. Erle Ellis et al., "Involve Social Scientists in Defining the Anthropocene," *Nature* 540, no. 7632 (2016): 192.

8. Elizabeth Kolbert, *The Sixth Extinction: An Unnatural History* (New York: Henry Holt, 2014).

9. Stuart L. Pimm et al., "The Biodiversity of Species and Their Rates of Extinction, Distribution, and Protection," *Science* 344, no. 6187 (2014): DOI: 10.1126/science. 1246752.

10. Alfred Crosby, *The Columbian Exchange: Biological and Cultural Consequences of 1492* (Westport, CT: Greenwood, 1972).

11. Daniel Simberloff and Michael Rejmanek, eds., *Encyclopedia of Biological Invasions* (Berkeley: University of California Press, 2011).

12. Erle C. Ellis and Navin Ramankutty, "Putting People in the Map: Anthropogenic Biomes of the World," *Frontiers in Ecology and the Environment* 6, no. 8 (2008): 439–47.

13. Peter Turchin, *War and Peace and War: The Rise and Fall of Empires* (New York: Penguin, 2007).

14. Charles Darwin, *The Descent of Man and Selection in Relation to Sex* (London: John Murray, 1871).

15. Frans de Waal, *The Age of Empathy: Nature's Lessons for a Kinder Society* (New York: Crown, 2010).

16. Sandel, *What Money Can't Buy.*

17. Robert Wokler, *Rousseau, the Age of Enlightenment, and Their Legacies* (Princeton, NJ: Princeton University Press, 2012).

18. Ara Norenzayan, *Big Gods: How Religion Transformed Cooperation and Conflict* (Princeton, NJ: Princeton University Press, 2013).

19. Jonathan Gottschall, *The Storytelling Animal: How Stories Make Us Human* (Boston: Houghton Mifflin Harcourt, 2012).

20. Roderick Frazier Nash, *Wilderness and the American Mind* (New Haven, CT: Yale University Press, 1982).

21. Robert M. Sapolsky, *Behave: The Biology of Humans at Our Best and Worst* (New York: Penguin, 2017).

22. The following sources provide further background reading: Kevin Kelly, *The Inevitable: Understanding the 12 Technological Forces That Will Shape Our Future* (New York: Penguin, 2017); Henry Jenkins, *Convergence Culture: Where Old and New Media Collide* (New York: New York University Press, 2006); Diane Ackerman, *The Human Age: The World Shaped by Us* (New York: Norton, 2014); Stephen A. Marglin, *The Dismal Science: How Thinking Like an Economist Undermines Community* (Cambridge, MA: Harvard University Press, 2008); Crosby, *The Columbian Exchange;* Richard Baldwin, *The Great Convergence: Information Technology and the New Globalization* (Cambridge, MA: Harvard University Press, 2016).

15. The Modern Conservation Movement

1. David Western and Wesley Henry, "Economics and Conservation in Third World National Parks," *BioScience* 29, no. 7 (1979): 414–18.

2. Alfred Runte, *National Parks: The American Experience* (Lincoln: University of Nebraska Press, 1979).

3. William Hornaday, cited in Runte, *National Parks.*

4. Noel Simon, *Between the Sunlight and the Thunder: The Wild Life of Kenya* (London: Collins, 1962).

5. Simon, *Between the Sunlight and the Thunder.*

6. Chase Alston, *Playing God in Yellowstone: The Destruction of America's First National Park* (Boston: Houghton Mifflin, 1987).

7. Thomas Michael Power and Richard Barrett, *Post-Cowboy Economics: Pay and Prosperity in the New American West* (Washington, DC: Island, 2001).

8. Robert B. Keiter, *To Conserve Unimpaired: The Evolution of the National Park Idea* (Washington, DC: Island, 2013).

9. Roderick Frazier Nash, *Wilderness and the American Mind* (New Haven, CT: Yale University Press, 1982).

10. Keith Thomas, *Man and the Natural World: A History of the Modern Sensibility* (New York: Pantheon Books, 1983).

11. Ralph Waldo Emerson, *Ralph Waldo Emerson: Selected Essays, Lectures, and Poems* (New York: Bantam Classics, 1990).

12. Henry David Thoreau, *Walden; or, Life in the Woods* (Boston: Ticknor & Fields, 1854).

13. Eric Rutkow, *American Canopy: Trees, Forests, and the Making of a Nation* (New York: Scribner, 2013).

14. Andrea Wulf, *The Invention of Nature: Alexander Von Humboldt's New World* (New York: Vintage, 2016).

15. George Perkins Marsh, *Man and Nature; or, Physical Geography as Modified by Human Action* (New York: Charles Scribner, 1864).

16. James A. Tober, *Who Owns the Wildlife? The Political Economy of Conservation in Nineteenth-Century America* (Santa Barbara, CA: ABC-CLIO, 1981).

17. Samuel P. Hays, *Conservation and the Gospel of Efficiency: The Progressive Conservation Movement, 1890–1920* (Pittsburgh: University of Pittsburgh Press, 1999).

18. Adam Smith, *The Wealth of Nations* (New York: Bantam Dell, 2003).

19. John F. Ross, *The Promise of the Grand Canyon: John Wesley Powell's Perilous Journey and His Vision for the Amercan West* (New York: Viking, 2018).

20. Hannah Holleman, *Dust Bowls of Empire: Imperialism, Environmental Politics, and the Injustice of Green Capitalism* (New Haven, CT: Yale University Press, 2018).

21. Andrew C. Isenberg, *The Destruction of the Bison: An Environmental History, 1750–1920* (Cambridge: Cambridge University Press, 2000).

22. Donald Worster, *Dust Bowl: The Southern Plains in the 1930s* (Oxford: Oxford University Press, 2004).

23. Holleman, *Dust Bowls of Empire*.

24. Timothy Egan, *The Worst Hard Time: The Untold Story of Those Who Survived the Great American Dust Bowl* (Boston: Houghton Mifflin, 2006).

25. Worster, *Dust Bowl*.

26. Jon Mooallem, *Wild Ones: A Sometimes Dismaying, Weirdly Reassuring Story about Looking at People Looking at Animals in America* (New York: Penguin, 2014).

27. Aldo Leopold, *A Sand County Almanac and Sketches Here and There* (Oxford: Oxford University Press, 1949).

28. Thomas Talhelm et al., "Large-Scale Psychological Differences within China Explained by Rice versus Wheat Agriculture," *Science* 344, no. 6184 (2014): 603–8.

29. Rachel Carson, *Silent Spring* (Boston: Houghton Mifflin, 1962).

30. Barry Commoner, *The Closing Circle: Nature, Man and Technology* (New York: Knopf, 1971).

31. Keiter, *To Conserve Unimpaired*.

16. Unnatural Reconnections

1. United Nations Department of Economics and Social Affairs Population Division, *World Population Prospects 2019*, vol. 1, *Comprehensive Tables*, /ST/ESA/SER A/426.

2. Joel K. Bourne, *The End of Plenty: The Race to Feed a Crowded World* (New York: Norton, 2015).

3. Ernesto Zedillo, *Global Warming: Looking beyond Kyoto* (Washington, DC: Brookings Institution Press, 2008); David Wallace-Wells, *The Uninhabitable Earth: Life After Warming* (New York: Tim Duggan Books, 2019).

4. Charles G. Curtin, *The Science of Open Spaces: Theory and Practice for Conserving Large, Complex Systems* (Washington, DC: Island, 2015).

5. Emma Marris, *Rambunctious Garden: Saving Nature in a Post-Wild World* (New York: Bloomsbury, 2013).

6. Stewart Brand, *Whole Earth Discipline: An Ecopragmatist Manifesto* (New York: Viking, 2009).

7. "2015 Edelman Trust Barometer," January 19, 2015, https://www.edelman.com/research/2015-edelman-trust-barometer.

8. Anthony D. Barnosky and Elizabeth A. Hadly, *Tipping Point for Planet Earth: How Close Are We to the Edge?* (Glasgow: William Collins, 2016).

9. Nicholas Stern, *The Economics of Climate Change: The Stern Review* (Cambridge: Cambridge University Press, 2007).

10. Fred Pearce, "The Green Shoots of Hope?" *New Scientist* 227, no. 3028 (2015): 42–43.

11. Robert Olson and David Rejeski, eds., *Environmentalism and the Technologies of Tomorrow: Shaping the Next Industrial Revolution* (Washington, DC: Island, 2004).

12. Robert Wright, *Nonzero: History, Evolution & Human Cooperation* (London: Abacus, 2001).

13. Don Tapscott and Anthony D. Williams, *Wikinomics: How Mass Collaboration Changes Everything* (London: Atlantic Books, 2007).

14. Howard Rheingold, *Smart Mobs: The Next Social Revolution* (New York: Basic Books, 2003).

15. Neil Gershenfeld and J. P. Vasseur, "As Objects Go Online: The Promise (and Pitfalls) of the Internet of Things," *Foreign Affairs,* April 2014.

16. Vannevar Bush, "As We May Think," *Atlantic Monthly,* July 1945.

17. Andy Clark, "Natural-Born Cyborgs?" in *International Conference on Cognitive Technology,* ed. Erik M. Altmann et al. (Berlin: Springer, 2001), 17–24.

18. Klaus Schwab, "The Fourth Industrial Revolution: What It Means and How to Respond," in *The Fourth Industrial Revolution: A Davos Reader,* ed. Gideon Rose (New York: Council on Foriegn Relations, 2016).

19. Daniela Rus, "The Robots Are Coming: How Technological Breakthroughs Will Transform Everyday Life," *Foreign Affairs,* July 2015.

20. Walter R. Stahel, "The Circular Economy," *Nature* 531, no. 7595 (2016): 435–38.

21. Schwab, "The Fourth Industrial Revolution"; Juho Hamari, Mimmi Sjöklint, and Antti Ukkonen, "The Sharing Economy: Why People Participate in Collaborative Consumption," *Journal of the Association for Information Science and Technology* 67, no. 9 (2016): 2047–59.

22. Rus, "The Robots Are Coming."

23. Mark Weiser, quoted in Laurie Garrett, "Biology's Brave New World: The Promise and Perils of the Synbio Revolution," in Rose, *The Fourth Industrial Revolution*.

24. Laurie Garrett, "Biology's Brave New World: The Promise and Perils of the Synbio Revolution," *Foreign Affairs*, November 2013.

25. Jon Gether, "Why Isn't the Brain Green?" *New York Times*, April 19, 2009.

26. Gether, "Why Isn't the Brain Green?"

27. Richard H. Thaler and Cass R. Sunstein, *Nudge: Improving Decisions about Health, Wealth, and Happiness* (New York: Penguin, 2009).

28. David Brooks, *The Social Animal: The Hidden Sources of Love, Character, and Achievement* (New York: Random House, 2012).

17. New Tools for a New Age

1. Hans Rosling, Ola Rosling, and Anna Rosling Ronnlund, *Factfulness: Ten Reasons We're Wrong about the World—and Why Things Are Better Than You Think* (New York: Flatiron Books, 2018).

2. Corey J. A. Bradshaw and Paul R. Ehrlich, *Killing the Koala and Poisoning the Prairie: Australia, America, and the Environment* (Chicago: University of Chicago Press, 2015).

3. Matt Ridley, *The Evolution of Everything: How New Ideas Emerge* (New York: Harper, 2015).

4. Joel E. Cohen, *How Many People Can the Earth Support?* (New York: Norton, 1996).

5. Vaclav Smil, *Feeding the World: A Challenge for the Twenty-First Century* (Cambridge, MA: MIT Press, 2000).

6. Rosling, Rosling, and Ronnlund, *Factfulness*.

7. David Griggs et al., "Policy: Sustainable Development Goals for People and Planet," *Nature* 495, no. 7441 (2013): 305.

8. Charles Fishman, *The Big Thirst: The Secret Life and Turbulent Future of Water* (New York: Free Press, 2012).

9. Philip Ball, *The Water Kingdom: A Secret History of China* (Chicago: University of Chicago Press, 2017).

10. Fishman, *The Big Thirst*.

11. John Fleck, *Water Is for Fighting Over: And Other Myths about Water in the West* (Washington, DC: Island, 2016).

12. Fishman, *The Big Thirst*.

13. Xin Zhang et al., "Managing Nitrogen for Sustainable Development," *Nature* 528, no. 7580 (2015): 51.

14. Ben Phalan et al., "Reconciling Food Production and Biodiversity Conservation: Land Sharing and Land Sparing Compared," *Science* 333, no. 6047 (2011): 1289–91.

15. Joseph Poore, "Back to the Wild," *New Scientist* 235, no. 3138 (2017): 26–29.

16. Helen Anne Curry, *Evolution Made to Order: Plant Breeding and Technological Innovation in Twentieth-Century America* (Chicago: University of Chicago Press, 2016).

17. Emma Marris, "Agronomy: Five Crop Researchers Who Could Change the World," *Nature* 456, no. 7222 (2008): 563–68.

18. Jennifer A. Doudna and Samuel H. Sternberg, *A Crack in Creation: Gene Editing and the Unthinkable Power to Control Evolution* (Boston: Houghton Mifflin Harcourt, 2017).

19. Michael Le Page, "Gene-Silencing Farming Revolution," *New Scientist*, no. 3108 (2017): 8.

20. Ottmar Edenhofer, "King Coal and the Queen of Subsidies," *Science* 349, no. 6254 (2015): 1286–87.

21. World Bank and the Institute for Health Metrics and Evaluation, *The Cost of Air Pollution: Strengthening the Economic Case for Action*, 2016, http://documents.worldbank.org/curated/en/781521473177013155/pdf/108141-REVISED-Cost-of-PollutionWebCORRECTEDfile.pdf.

22. "A Renewable Energy Boom," *New York Times*, April 4, 2016.

23. Walter R. Stahel, "The Circular Economy," *Nature News* 531, no. 7595 (2016): 435.

24. Steve Hatfield-Dodds et al., "Australia Is 'Free to Choose' Economic Growth and Falling Environmental Pressures," *Nature* 527, no. 7576 (2015): 49.

18. Cleaning Our Planetary Nest

1. Emily Underwood, "The Polluted Brain," *Science* 27355, no. 6323 (2017): 342–45.

2. Thomas Dietz et al., *New Tools for Environmental Protection: Education, Information, and Voluntary Measures* (Washington, DC: National Academies Press, 2002).

3. Ewen Callaway, "Daily Dose of Toxics to Be Tracked," *Nature* 491, no. 7426 (2012): 647.

4. Leonardo Trasande, "Stand Firm on Hormone Disruptors," *Nature* 539, no. 7630 (2016): 469.

5. Michael Pollan, *In Defence of Food: The Myth of Nutrition and the Pleasures of Eating* (London: Allen Lane, 2008).

19. Nature and Human Well-being

1. Charles Darwin, *The Descent of Man and Selection in Relation to Sex* (London: John Murray, 1871).

2. Amotz Zahavi and Avishag Zahavi, *The Handicap Principle: A Missing Piece of Darwin's Puzzle* (Oxford: Oxford University Press, 1997).

3. David Rothenberg, *Survival of the Beautiful: Art, Science, and Evolution* (London: Bloomsbury, 2011).

4. Richard O. Prum, *The Evolution of Beauty: How Darwin's Forgotten Theory of Mate Choice Shapes the Animal World—and Us* (New York: Doubleday, 2017).

5. Stephen Jay Gould and Richard C. Lewontin, "The Spandrels of San Marco and the Panglossian Paradigm: A Critique of the Adaptationist Programme," *Proceedings of the Royal Society of London: Series B, Biological Sciences* 205, no. 1161 (1979): 581–98.

6. World Travel and Tourism Council, "Travel and Tourism Economic Impact," 2014, www.wttc.org.

7. Outdoor Industry Association, "The Outdoor Recreation Economy," 2017, outdoorindustry.org.

8. Keith Thomas, *Man and the Natural World: A History of the Modern Sensibility* (New York: Pantheon Books, 1983).

9. Cynthia Moss, *Echo of the Elephants: The Story of an Elephant Family* (New York: William Morrow, 1993).

10. Richard Louv, *Last Child in the Woods: Saving Our Children from Nature-Deficit Disorder* (Chapel Hill, NC: Algonquin Books, 2006); Richard Louv, *The Nature Principle: Reconnecting with Life in a Virtual Age* (Chapel Hill, NC: Algonquin Books, 2011).

11. Edward O. Wilson, *Half-Earth: Our Planet's Fight for Life* (New York: Liveright, 2016).

12. Roderick Frazier Nash, *Wilderness and the American Mind* (New Haven, CT: Yale University Press, 1982).

20. Natural Capital

1. David Western, "An African Odyssey to Save the Elephant," *Discover* 7, no. 10 (1986): 56–70.

2. David Western, "A Half a Century of Habitat Change in Amboseli National Park, Kenya," *African Journal of Ecology* 45, no. 3 (2007): 302–10.

3. John Waithaka, "The Ecological Role of Elephants in Restructuring Plant and Animal Communities in Different Eco-Climatic Zones in Kenya" (PhD diss., Kenyatta University, 1994).

4. Norman Owen-Smith, "Megafaunal Extinctions: The Conservation Message from 11,000 Years BP," *Conservation Biology* 3, no. 4 (1989): 405–12.

5. Chase Alston, *Playing God in Yellowstone: The Destruction of America's First National Park* (Boston: Houghton Mifflin, 1987).

6. Robert T. Paine, "A Note on Trophic Complexity and Community Stability," *American Naturalist* 103, no. 929 (1969): 91–93.

7. John Terborgh and James A. Estes, eds., *Trophic Cascades: Predators, Prey, and the Changing Dynamics of Nature* (Washington, DC: Island, 2010).

8. Johan T. du Toit, Kevin H. Rogers, and Harry C. Biggs, eds., *The Kruger Experience: Ecology and Management of Savanna Heterogeneity* (Washington, DC: Island, 2003).

9. Paul Schultz Martin, *Twilight of the Mammoths: Ice Age Extinctions and the Rewilding of America* (Berkeley: University of California Press, 2005).

10. Jon Mooallem, *Wild Ones: A Sometimes Dismaying, Weirdly Reassuring Story about Looking at People Looking at Animals in America* (New York: Penguin, 2014).

11. Cristina Eisenberg, *The Wolf's Tooth: Keystone Predators, Trophic Cascades, and Biodiversity* (Washington, DC: Island, 2013).

12. Aldo Leopold, *A Sand County Almanac and Sketches Here and There* (Oxford: Oxford University Press, 1949).

13. James E. Lovelock, *Gaia: A New Look at Life on Earth* (Oxford: Oxford University Press, 1979); Lawrence E. Joseph, *Gaia—The Growth of an Idea* (New York: St. Martin's, 1990).

14. David Western and Anna K. Behrensmeyer, "Bone Assemblages Track Animal Community Structure over 40 Years in an African Savanna Ecosystem," *Science* 324, no. 5930 (2009): 1061–64.

15. Ken Thompson, *Where Do Camels Belong? The Story and Science of Invasive Species* (London: Profile Books, 2014).

16. Michelle Nijhuis, "How the Parks of Tomorrow Will Be Different," *National Geographic*, December 2016.

17. Peter Ward, *The Media Hypothesis: Is Life on Earth Ultimately Self-Destructive?* (Princeton, NJ: Princeton University Press, 2009).

18. Daniel Botkin, *Discordant Harmonies: A New Ecology for the Twenty-First Century* (Oxford: Oxford University Press, 1990).

19. David Pimentel and Michael Burgess, "Environmental and Economic Costs of the Application of Pesticides Primarily in the United States," in *Integrated Pest Management,* ed. Rajinder Pershin and David Pimentel (New York: Springer, 2014), 47–71.

20. Jim Sterba, *Nature Wars: The Incredible Story of How Wildlife Comebacks Turned Backyards into Battlegrounds* (New York: Crown, 2012).

21. David Cuff and Andrew S. Goudie, eds., *Encyclopaedia of Global Change* (Oxford, Oxford University Press, 2001).

22. World Wide Fund for Nature, "Living Planet Report 2016: Risk and Resilience in a New Era," 2016, https://www.worldwildlife.org/pages/living-planet-report-2016.

23. David Western, Samantha Russell, and Innes Cuthill, "The Status of Wildlife in Protected Areas Compared to Non-protected Areas of Kenya," *PloS One* 4, no. 7 (2009): e6140.

24. Convention on Biological Diversity, *Global Biodiversity Outlook 3,* 2010, https://www.cbd.int/doc/publications/gbo/gbo3-final-en.pdf.

25. Gretchen C. Daily, ed., *Nature's Services: Societal Dependence on Natural Ecosystems,* 2nd ed. (Washington, DC: Island, 1997).

26. Partha Dasgupta, *Human Well-Being and the Natural Environment* (Oxford: Oxford University Press, 2001).

27. Robert Costanza et al., "Changes in the Global Value of Ecosystem Services," *Global Environmental Change* 26 (2014): 152–58.

28. Paulo A. L. Nunes, Pushpam Kumar, and Tom Dedeurwaedere, eds., *Handbook on the Economics of Ecosystem Services and Biodiversity* (Cheltenham, UK: Edward Elgar, 2014).

29. *Kenya's Natural Capital: A Biodiversity Atlas* (Nairobi: Government of Kenya, 2015).

30. The following sources provide further background reading: William E. Odum, "Environmental Degradation and the Tyranny of Small Decisions," *BioScience* 32, no. 9 (1982): 728–29; Millennium Assessment Board, *Millennium Ecosystem Assessment* (Washington, DC: Island, 2005); Alfred E. Kahn, "The Tyranny of Small Decisions:

Market Failures, Imperfections, and the Limits of Economics," *Kyklos* 19, no. 1 (1966): 23–47; Thomas Prugh et al., *Natural Capital and Human Economic Survival* (Oxford: Oxford University Press, 2011); Pushpam Kumar, ed., *The Economics of Ecosystems and Biodiversity: Ecological and Economic Foundations* (New York: Routledge, 2010).

21. The City and the Planet

1. Carl Sagan, *The Varieties of Scientific Experience: A Personal View of the Search for God* (New York: Penguin, 2006).

2. Tim Flannery, *Here on Earth: A Twin Biography of the Planet and the Human Race* (London: Penguin, 2012).

3. Chris D. Thomas et al., "Extinction Risk from Climate Change," *Nature* 427, no. 6970 (2004): 145.

4. Tim Newbold et al., "Has Land Use Pushed Terrestrial Biodiversity beyond the Planetary Boundary? A Global Assessment," *Science* 353, no. 6296 (2016): 288–91.

5. Margaret Munro, "What's Killing the World's Shorebirds?" *Nature* 541, no. 7635 (2017): 16.

6. Elinor Ostrom, *Governing the Commons: The Evolution of Institutions for Collective Action* (Cambridge: Cambridge University Press, 1990).

7. Gernot Wagner and Martin L. Weitzman, *Climate Shock: The Economic Consequences of a Hotter Planet* (Princeton, NJ: Princeton University Press, 2015).

8. William McDonough, "Carbon Is Not the Enemy," *Nature,* November 14, 2016, 349–51.

9. Norman Myers and Jennifer Kent, *Perverse Subsidies: How Tax Dollars Can Undercut the Environment and the Economy* (Washington, DC: Island, 2001).

10. Tara Martin, "Policy: Hasten End of Dated Fossil-Fuel Subsidies," *Nature* 538, no. 7624 (2016): 171.

11. William D. Nordhaus, *The Climate Casino: Risk, Uncertainty, and Economics for a Warming World* (New Haven, CT: Yale University Press, 2013).

12. Stewart Brand, *Whole Earth Discipline: An Ecopragmatist Manifesto* (New York: Viking, 2009).

13. Luís M. A. Bettencourt et al., "Growth, Innovation, Scaling, and the Pace of Life in Cities," *Proceedings of the National Academy of Sciences* 104, no. 17 (2007): 7301–6.

14. Marcus J. Hamilton et al., "Nonlinear Scaling of Space Use in Human Hunter-Gatherers," *Proceedings of the National Academy of Sciences* 104, no. 11 (2007): 4765–69.

15. Geoffrey B. West, *Scale: The Universal Laws of Growth, Innovation, Sustainability, and the Pace of Life in Organisms, Cities, Economies, and Companies* (New York: Penguin, 2017).

16. Duncan McLaren and Julian Agyeman, *Sharing Cities: A Case for Truly Smart and Sustainable Cities* (Cambridge, MA: MIT Press, 2015).

17. West, *Scale.*

18. Flannery, *Here on Earth.*

19. Jane Jacobs, *The Economies of Cities* (New York: Vintage Books, 1970).

20. Edward L. Glaeser, *Triumph of the City: How Our Greatest Invention Makes Us Richer, Smarter, Greener, Healthier, and Happier* (New York: Penguin, 2011).

21. Nicholas S. Wigginton et al., "Cities Are the Future," *Science* 352, no. 6288 (2016): 1904–5.

Conclusion

1. Gary Cohen, "America Is Competitive Again," *Wall Street Journal,* December 20, 2018.

2. Michael R. Bloomberg, "Climate Progress, with or without Trump," *New York Times,* March 31, 2017.

3. "As the U.S. Sheds Role as Climate Change Leader, Who Will Fill the Void?" *New York Times,* November 17, 2017.

4. James C. McCann, *Maize and Grace* (Cambridge, MA: Harvard University Press, 2005).

5. Joshua David Greene, *Moral Tribes: Emotion, Reason, and the Gap between Us and Them* (New York: Penguin, 2013).

6. Peter Singer, *The Expanding Circle: Ethics and Sociobiology* (New York: Farrar, Straus & Giroux, 1981).

7. Bryan G. Norton, *Toward Unity among Environmentalists* (Oxford: Oxford University Press, 1991).

8. The following sources provide further background reading: Christine E. Gudorf and James E. Huchingson, *Boundaries: A Casebook in Environmental Ethics* (Washington, DC: Georgetown University Press, 2003); Susan Clayton and Gene Myers, *Conservation Psychology: Understanding and Promoting Human Care for Nature* (New York: Blackwell, 2009); Ronald J. Engel and Joan Gibb Engel, eds., *Ethics of Environment and Development: Global Challenge, International Response* (Tucson: University of Arizona Press, 1990).

Acknowledgments

I have previously acknowledged in papers and books the many sponsors, affiliates, and friends who have contributed to my conservation and research work over the last fifty years.

Here I wish to thank those who have contributed informally to the views I express in *We Alone*. Most especially, I thank the communities in Tanzania, Kenya, and elsewhere who have generously shared with me their skills and knowledge of surviving harsh times and thriving alongside wildlife for millennia. Except for them, Africa's wealth of wildlife would have vanished as surely as it has on other continents. The lessons I've learned firsthand from subsistence communities sparked my fascination with the role of conservation in our rise from our primate ancestry in the savannas to global dominance and the ultimate challenges of living within planetary limits.

Special thanks to Parashino Ole Purdul, Daniel Sindiyo, Kerenkol Ole Musa, Shuhe Mweyendet, Daniel Ole Somoire, Jonathan Lebo, John Marinka, Koikai Ole Tiptip, Benson Laiyan, John Kamanga, Johnson Sipitiek, and Jackson Mwato. I owe a special gratitude to my field research assistant of over forty years, David Maitumo.

Helen Gichohi, John Waithaka, Chris Gakahu, Lucy Waruingi, Victor Mose, Samantha Russell, and the many students, conservationists, and researchers at the African Conservation Centre have been an inspiration, spurring my conservation work and thinking over the years. Bill Conway, long-term president of the Wildlife Conservation Society (WCS), gave me full rein to follow my nose, often in directions running counter to the conservation philosophy and practices of the day. Also at WCS, George Schaller, Chuck Carr, Roger Payne, Tom Struhsaker, Alan Rabinowitz, Amy Vedder, and Bill Weber inspired me with their dedication to saving species and wild places around the world.

I also wish to thank Liz Claiborne and Art Ortenberg, who did so much to make community-based conservation a reality around the world, and David Quammen, Mary Pearl, Alison and Bob Richard, Bill Conway, George Schaller, Bill de Buys, Doug

Chadwick, Guy Grant, Jim Murtaugh, and Victor Kovner, who shared in the vision and mission of the Liz Claiborne & Art Ortenberg Foundation in promoting coexistence between people and wildlife.

Among my academic colleagues I have benefited from decades of exchanges with Tom Dunne above all, and also David Woodruff, Nick Georgiadis, James Ssemakula, Chris Gakahu, Virginia Finch, Steve Cobb, Harvey Croze, Cynthia Moss, Frank Mitchell, Bruno Latour, Debbie Nightingale, Charis Cussins, Charles Curtin, Rick Potts, Kay Behrensmeyer, Norman Myers, Walde Tadesse, Marissa Ahlering, Walter Jetz, Samantha Russell, Peadar Brehoni, Pete Tyrell, and many others whom I've had the pleasure of joining in workshops, conferences, and field trips around the world.

I express my great appreciation to friends and colleagues with whom I've worked on my various assignments with the Kenya government, especially the wardens, rangers, researchers, administrators, permanent secretaries, ministers, civil servants, and politicians who have influenced my views on the importance of national governance and global agreements.

My father was the most important influence in my early years, first on the hunting trail and later in saving rather than killing animals. My mother Bea's love of animals fired up my own passion for wildlife. My sibs, Sheila, Martin, and Lynne, were a part of my long journey from bush kid to conservationist.

Jonathan Cobb has been the finest of editors in offering advice on the first draft of the book and editing the two further drafts. Victor Mose, Winnfridah Kemunto, and Erastus Mwniki helped compile the photos. Very special thanks to my agent Laura Black Peterson for steering the book to publication, and to Jean Thomson Black, senior executive editor at Yale University Press, for taking on the book assignment. Robin DuBlanc did a splendid job copyediting the book. I also owe special thanks to Peter Raven and an anonymous reviewer for their comments and advice for improving the book.

Finally, thanks beyond words go to my wife, Shirley Strum, and our children, Carissa and Guy, for their enormous forbearance and support over the years, often during hard times and tough assignments. I relish all those wonderful safaris we shared in Amboseli among the elephants, in Chololo among the baboons, and around and beyond Kenya. Shirley has been the biggest supporter and firmest critic of my conservation and research work over our years together and from the inception to the completion of *We Alone*. I only hope my support for her next book is half as helpful.

Index